ANTENTOP

ANTENTOP 01 2015 # 019

ANTENTOP is *FREE* e-magazine devoted to ANTENna's
Theory,
Operation, and
Practice

1-2015

Edited by hams for hams

In the Issue:
Antennas Theory!

Practical design of HF Antennas!

Antenna Transformers!

Practical design of 50 and 70- MHz Antennas!

Regenerative Receivers!

Antenna for 50 and 70- MHz Band by ES4UY (also UY5ON)

Thanks to our authors:

Prof. Natalia K.Nikolova

Nick Kudryavchenko, UR0GT

Aleksandr Grachev, UA6AGW

Vasily Perov, DL1BA

And others…..

DL1BA Symmetrical Antenna Tuner

EDITORIAL:
Well, my friends, new ANTENTOP – 01 - 2015 come in! ANTENTOP is just authors' opinions in the world of amateur radio. A little note, I am not native English, so, of course, there are some sentence and grammatical mistakes there… Please, be indulgent!
ANTENTOP 01 –2015 contains antenna articles, description of antenna RX and TX transformers, Regenerative Receivers. Hope it will be interesting for you.
Our pages are opened for all amateurs, so, you are welcomed always, both as a reader as a writer.

Copyright: Here at ANTENTOP we just wanted to follow traditions of FREE flow of information in our great radio hobby around the world. *A whole issue* of ANTENTOP may be photocopied, printed, pasted onto websites. We don't want to control this process. It comes from all of us, and thus it belongs to all of us. This doesn't mean that there are no copyrights.

There is! *Any work is copyrighted by the author. All rights to a particular work are reserved by the author.*

73! Igor Grigorov, VA3ZNW

ex: RK3ZK, UA3-117-386,

UA3ZNW, UA3ZNW/UA1N, UZ3ZK

op: UK3ZAM, UK5LAP, EN1NWB, EN5QRP, EN100GM

Contact us: Just email me
igor.grigorov@gmail.com

NB: Please, use only plain text and mark email subject as: **igor_ant**. I receive lots spam, so, I delete **ALL** unknown me messages **without** reading.

ANTENTOP is *FREE* **e-magazine, available** *FREE* **at** http://www.antentop.org/

ANTENTOP- 01- 2015, # 019 — Editorial

Welcome to ANTENTOP, FREE e - magazine!

ANTENTOP is *FREE e- magazine*, made in *PDF*, devoted to Antennas and Amateur Radio. Everyone may share his experience with others hams on the pages. Your opinions and articles are published without any changes, as I know, every your word has the mean.

Every issue of ANTENTOP is going to have 100 pages and this one will be paste in whole on the site. I do not know what a term for one issue would be taken, may be 12 month or so. A whole issue of ANTENTOP holds nearly 10- 30 MB.

A little note, I am not native English, so, of course, there are some sentence and grammatical mistakes there... Please, be indulgent!

Publishing: If you have something for share with your friends, and if you want to do it *FREE*, just send me an email. Also, if you want to offer for publishing any stuff from your website, you are welcome!

Your opinion is important for me, so, contact if you want to say something!

Copyright Note:

Dear friends, please, note, I respect Copyright. Always, when I want to use some stuff for ANTENTOP, I ask owners about it. But... sometimes my efforts have no success. I have some very interesting stuff from closed websites however their owners keep silence... as well as I have no response on some my emails from some owners.

I have a big collection of pictures. I have got the pictures and stuff in different ways, from *FREE websites*, from commercial CDs, intended for *FREE using*, and so on... I use to the pictures (and seldom, some stuff from free and closed websites) in ANTENTOP. *If the owners of the Copyright stuff have concern*, please, contact with me, I immediately remove any Copyright stuff, or, if it is necessary, all needed references will be made there.

Business Advertising: **ANTENTOP is not a commercial magazine.** Authors and I (Igor Grigorov, the editor of the magazine) do not get any profit from any issue. But of course, I do not mention from commercial ads in ANTENTOP. It allows me to do the magazine in most great way, allows me to pay some money for authors to compensate their hard work.

So, if you want paste a commercial advertisement in ANTENTOP, please contact me.

Book Advertising: I believe that *Book Advertising* is a noncommercial advertisement. So, Book Advertising is *FREE* at ANTENTOP. Contact with me for details.

Email: igor.grigorov@gmail.com
subject: *igor_ant*

NB: Please, use only plain text and mark email subject as: *igor_ant*. I receive lots spam and viruses, so, I delete *ALL unknown me messages* without reading.

73! *Igor Grigorov*, VA3ZNW
ex: UA3-117-386, UA3ZNW, UA3ZNW/UA1N, UZ3ZK, RK3ZK
op: UK3ZAM, UK5LAP, EN1NWB, EN5QRP, EN100GM

http://www.antentop.org/

Table of Contents

Antenna Theory

Page

Linear Array Theory- Part III : by: Prof. Natalia K. Nikolov

1. Dear friends, I would like to give to you an interesting and reliable antenna theory. Hours searching in the web gave me lots theoretical information about antennas. Really, at first I did not know what information chose for ANTENTOP.

   Now I want to present to you one more very interesting Lecture - it is LECTURE 17: Linear Array Theory- Part III : N-element linear array with uniform spacing and non-uniform amplitude: binomial array; Dolph-Tschebyscheff array; directivity and design considerations......

   5- 26

HF- Antenna Practice

Two Slopers for All Traditional Five HF-Bands: By Vladimir Fursenko, UA6CA

2. It is possible at one mast install two slopers that cover all traditional five HF-Bands- 80,- 40,- 40,- 15 and 10- meter.......

   27

Antenna for 80-, 40-, 20-, 17-, 15-, 12-, and 10- meter HF Band: By Vladimir Fursenko, UA6CA

3. The two- level antenna works at 80-, 40-, 20-, 17-, 15-, 12- and 10- meter Bands. Upper level works on 80-, 40-, 20-, 15- and 10- meter Bands. Lower level works on 17- and 12- meter Bands. ...

   28- 29

Bidirectional Vertical Antenna for the 20- meter Band: By: Nikolay Kudryavchenko, UR0GT

4. One Ground Plane allows get one directional switchable DD. For this purpose to the GP a two wires should be added. With help an RF Relay (or just with help of manually installed jumper) the wires turn on to Director and Reflector.....

   30- 31

L-Vertical Antenna for Nearest Objects for The 40- and 20- meter Bands: By: Nikolay Kudryavchenko, UR0GT

5. One L-Vertical Antenna may works at the two amateur Bands- 40- and 20- meters. The main feature of the antenna is that it may be located at nearest conductive objects- for example the antenna may be placed near water drain. Antenna may be placed at a wall of a house. Inside a house there are always a lot of conductive objects- for example electrical main, refrigerator, metal tubes of house ventilation system and so on.......

   32- 34

Table of Contents

Page		Page
6	**Universal Beverage Antenna:** by: Igor Grigorov, va3znw, Richmond Hill, Canada	35- 41
	Beverage Antennas are widely used at commercial and military radio communication. In commercial communication Beverage Antenna as usual is used as a receiving antenna. However, in military communication Beverage Antenna is used for both purposes- for receiving and transmitting applications. Transmitting/receiving Beverage Antenna was used in DX- Pedition EK1NWB on to Kizhy island where the antenna (against skepticism of some persons) illuminated its good job. So when again in Toronto I have changed my QTH and the QTH allowed me install Beverage Antenna, I did not hesitated...	
7	**Windom Compendium from RZ9CJ:** By: Sergey Popov, RZ9CJ, Ekaterinburg, Russia	42
	Just description of five Windom Antenas....	
8	**Windom UR0GT:** by: By: Nikolay Kudryavchenko, UR0GT	43
	Windom is one of the oldest and reliable antennas that used in ham radio. There are lots modifications of Windom Antenna (or in other words Off Center Dipole Antenna). One of such modification was optimized by UR0GT. The antenna was optimized for 40, 20 and 10- meter bands. ..	
9	**Two Vertical Antennas for 20-, 15- and 10- meter Bands:** By: Nikolay Kudryavchenko, UR0GT	44- 49
	Below described two vertical antennas that work without any ATU at the 20-, 15 and 10- meter Bands. The antennas easy to made and easy to tune to the bands.....	
10	**R3PIN Experimenters with UA6AGW Antenna** By: Aleksandr Grachev, UA6AGW Credit Line: CQ-QRP # 48 (Autumn 2014), pp.: 19- 22.	50- 52
	There are below described experimenters with UA6AGW Antenna made by Sergey Tetuyhin, R3PIN. Sergey would like create an UA6AGW Antenna for 2- meter Band. He did not have schematic of the antenna for 2- meter Band. He made two antennas that he believed would work at the 2- meter Band. However his attempt was not successful. But Sergey during the experimenters found some unusual sides at UA6AGW Antenna.	

Table of Contents

		Page
11	**Dipole Antenna for 40- and 20- meter Bands: By: Vasily Perov, DL1BA (ex UK8BA) Credit Line: Forum from: www.cqham.ru**	53- 55
	Because I have no lots space at my backyard for antenna installation I like do experiments with shortened antennas. Below described one of my experimental shortened dipole antenna for 40- and 20- meter Bands. It takes for me only 1 and half hour for installation and tuning of the antenna. After that the dipole antenna was tested at CQ WW Contest (2015). I made 300 QSOs using 100- Wtts going into the antenna....	
12	**Modified Dipole Antenna DL1BA for 40- and 20- meter Bands: By: Igor Vakhreev, RW4HFN**	56- 57
	In my opinion the explanation how the DL1BA Antenna (Antentop 01- 2015, pp: 53-55, Dipole Antenna for 40- and 20- meter Bands) is working at the 20- meter Band is very simple. Parts of the antenna- there are long wire (5.6-meter length) before inductor, inductor and short wire (1.5- meter length) after inductor - make 1.5- lambda dipole at the 20- meter Band.......	
13	**Modified DL1BA Dipole Antenna for 40- and 20- meter Bands with additional 10- or 15- meter Band: By: Igor Vakhreev, RW4HFN**	58
	DL1BA Antenna (Antentop 01- 2015, pp: 53-55, Dipole Antenna for 40- and 20- meter Bands) may be modified for working at additional 10- or 15-meter Band......	
14	**Modified DL1BA Dipole Antenna for 40-, 20-, 15-, and 10- meter Bands: By: Igor Vakhreev, RW4HFN**	59
	DL1BA Antenna (Antentop 01- 2015, pp: 53-55, Dipole Antenna for 40- and 20- meter Bands) may be modified for working at additional 10- and 15-meter Bands.....	
15	Antenna for 80-, 40-, and 15- meter Bands: By: Vladas Zhalnerauskas, UP2NV	60
	W3DZZ Antenna is widely used among radio amateurs. The antenna shows the efficiency at several bands (as usual from 3 to 5) at minimal stuff to do it. The described below antenna is a modification of the W3DZZ Antenna. The antenna could work at 80- 40- and 15- meter Bands.	

Find All Books from FREE ANTENTOP Amateur Library

Table of Contents

Page

HF- Antenna Practice

Antenna for 50 and 70- MHz Band: By: Alex Karakaptan, ES4UY, UY5ON

16 — The antenna was designed several years ago, when I got Estonian call sign ES4UY. With the call I able use 50 and 70- MHz bands from Estonia. Need to say that I was in a very rare square- KO49CJ. So I need an antenna for those bands. Restricted place could not allow me to create something serious. .... — 61-62

HF ATU

Symmetrical ATU: By: Vasily Perov, DL1BA (ex UK8BA)

17 — Prototype of the tuner was made by VK5RG. The tuner was found by me at "Das DARC Antennenbuch". However at the book there was given only brief description of the unit. The tuner takes my attention and by trial-and-error method I found the design (data for Inductors and Capacitors) of the tuner. .... — 63- 64

VHF ANTENNAS

Three Element Yagi Antenna for 145- MHz with Square Reflector: By: Yuriy Skutelis, RN3DEK

18 — The antenna provides good F/B ratio. Antenna has input impedance 50- Ohm that allows fed the antenna directly through 50- Ohm coaxial cable. It was reached by special form of the reflector. ... — 65- 66

Three Element Yagi Antenna for 145- MHz with Rectangle Reflector: By: Yuriy Skutelis, RN3DEK

19 — he antenna has F/B ratio at least 29 dB. It was reached by special form of the reflector. ... . — 67- 68

Four Element Antenna for Stack Design for 145- MHz Band By: Nikolay Kudryavchenko, UR0GT

20 — The four element YAGI is designed for installing in Four Element Stack Antenna System. The antennas not critical to nearest objects. Four such antennas are installed at corners of a quad.... — 69- 70

Table of Contents

Page

UHF ANTENNAS

21 Vertical Antenna 5/8 Lambda for 70- cm Band: By: Antonhax: Credit Line: Forum from: www.cqham.ru — 71- 72

At first the antenna was modeled by copper wire. Then the antenna was made on the base of bicycle wheel spokes. ....

22 **Broadband Vertical for 430- MHz Band: By: Nikolay Kudryavchenko, UR0GT** — 73- 74

Broadband Vertical Collinear Vertical antenna designed for 430- MHz Band. The antenna has Diagram Directivity with low-altitude maxima to the ground. Antenna has passband near 70- MHz at SWR 1.5:1.0. ....

TV ANTENNAS

23 **TV Antennas for Distance Receiving By: Leonid Pozdnyakov Credit Line: Radio # 10, 1953, pp.: 53- 54.** — 75- 77

At the original articles published at Radio # 10, 1953, there were described several antennas for distance receiving TV broadcasting stations. Below it is described one of those antennas- it is a Rhombic Antenna. Rhombic Antenna is easy to make and at the same time has perfect parameters...

RECEIVING ANTENNAS

24 **UB5UG Horizontal Receiving Antenna: By: Yuri Medinets, UB5UG** — 78- 79

At modern city to use a separate receiving antenna may be only one variant to be on the Air. Interferences from nearest electronics devices could force it. Below here it is described receiving antenna from far 70-s that may solve the hard modern situation.....

25 **Insulation RX Transformer: By: Igor Grigorov, va3znw** — 80- 82

At my shack I have used a Coaxial Antenna Switch Protax CSR- 5G to change devices switched to my antenna. The switch is very convenient for amateur operation in the Air. I can easy switch antenna from one transceiver (ICOM- 718) to another one (K1) or turn antenna to general coverage receiver (Hallicrafters S-85). I use the receiver to check propagation in the Air and just to catch some interesting HF- stations....

Table of Contents

Page

TUBE RECEIVERS

Simple Tube DC SSB Receiver: By R3KCR, Voronezh, Russia

26   It is simple experimental receiver that was made on two tubes- twin triodes. The receiver was tuned to 80- meter amateur band. 83- 84

HF Receiver for Beginner Ham: By: Viktor Lomanovich, UA3DH

27   85- 87

The HF Tube Regenerative Receiver is a classical design of the Tube Era. Somebody gave me magazine with the article at the 70s. I did the receiver and got very good result. I have received lots of amateur stations with the receiver. Then the receiver was remade by me in general coverage HF- Receiver. I could receive with great quality forbidden BBC, Voice of America lots broadcasting stations and easy found at those times "Numbers Stations." ...

BOOKS

28   **Direct Conversation Technique for Radio Amateurs. Vladimir Polyakov, RA3AAE, Ph. D in Technical Science**   88

Light description of the book and link to download....

29   **Underground and Ground Antennas: Georgiy A. Lavrov, Aleksey S. Knyazev Publishing House: Sovetskoe Radio, Moscow, 1965**   89

Light description of the book and link to download....

30   **Jones Antenna Handbook**   90

Light description of the book and link to download....

31   **Field Antenna Handbook**   91

Light description of the book and link to download....

32   **Construction of a Rhombic Receiving Antenna**   92

Light description of the book and link to download....

33   **Antennas and Antenna Systems**   93

Light description of the book and link to download....

34   **Antennas and Radio Propagation**   94

Light description of the book and link to download....

Table of Contents

#	Title	Page
35	**Design Handbook for High Frequency Radio Communications Systems** Light description of the book and link to download....	95

RF Transformers

#	Title	Page
36	**Broadband Transformer 50/200 Ohm: By: Sergey Popov, RZ9CJ, Ekaterinburg, Russia** Below I describe a simple way to make broadband transformer 50/200 Ohm with isolated windings. (Theoretically the transformer is for 50/140- Ohm. However it works fine for most common using 50/200 Ohm.)........	96- 97
37	**Two Broadband Symmetrical Transformers for HF and VHF Bands: By: Alex Karakaptan, UY5ON, ES4UY, Kharkov, Ukraine** For operation in the Air at all HF- Bands I use to antenna Delta. The antenna is fed by 300- Ohm Ladder Line. To match the antenna with my transceiver I use to ATU MFJ-962D. The ATU has symmetrical transformer at output. The transformer could provide good symmetrical operation ... but with antennas that has low reactance. My Delta has significant reactance through amateur's bands. So the concept is not for me....	98

Experimenters

#	Title	Page
38	**Experimenters with Microwave Oven: by Igor Grigorov, va3znw** Almost everyone has at home a Microwave Oven. It is possible make some experimenters with it. Most interesting and visual experiment is Experiment with Bulbs. We may find how microwaves effect to incandescent (filament) and CFL bulbs......	99- 100

Antenna Theory

www.antentop.org

Feel Yourself a Student!

Dear friends, I would like to give to you an interesting and reliable antenna theory. Hours searching in the web gave me lots theoretical information about antennas. Really, at first I did not know what information to choose for ANTENTOP. Finally, I stopped on lectures "Modern Antennas in Wireless Telecommunications" written by Prof. Natalia K. Nikolova from McMaster University, Hamilton, Canada.

You ask me: Why?

Well, I have read many textbooks on Antennas, both, as in Russian as in English. So, I have the possibility to compare different textbook, and I think, that the lectures give knowledge in antenna field in great way. Here first lecture "Introduction into Antenna Study" is here. Next issues of ANTENTOP will contain some other lectures.

So, feel yourself a student! Go to Antenna Studies!

I.G.

My Friends, the above placed Intro was given at ANTENTOP- 01- 2003 to Antennas Lectures.

Now I know, that the Lecture is one of popular topics of ANTENTOP. Every Antenna Lecture was downloaded more than 1000 times!

Now I want to present to you one more very interesting Lecture 17- it is a Lecture **Linear Array Theory- Part III**. *I believe, you cannot find such info anywhere for free! Very interesting and very useful info for every ham, for every radio- engineer.*

So, feel yourself a student! Go to Antenna Studies!

I.G.

McMaster University Hall

Prof. Natalia K. Nikolova

Linear Array Theory- Part III

N-element linear array with uniform spacing and non-uniform amplitude: binomial array; Dolph–Tschebyscheff array; directivity and design considerations...

by Prof. Natalia K. Nikolova

LECTURE 17: LINEAR ARRAYS – PART III
(*N-element linear array with uniform spacing and non-uniform amplitude: binomial array; Dolph–Tschebyscheff array; directivity and design considerations.*)

1. Advantages of linear array with non uniform amplitude
   The most often met BSAs, classifed according to the type of their excitation amplitude, are:
   a) the uniform BSA – relatively high directivity, but the side-lobe levels are high;
   b) Dolph–Tschebyscheff (Chebyshev, Чебышев) BSA – for a given number of elements directivity next after that of the uniform BSA, but side-lobe levels are the lowest in comparison with the other two types of arrays for a given directivity.
   c) Binomial BSA – does not have good directivity but has very low side-lobe levels (when $d = \lambda/2$, there are no side lobes at all).

2. Array factor (AF) of a linear array with non-uniform amplitude distribution

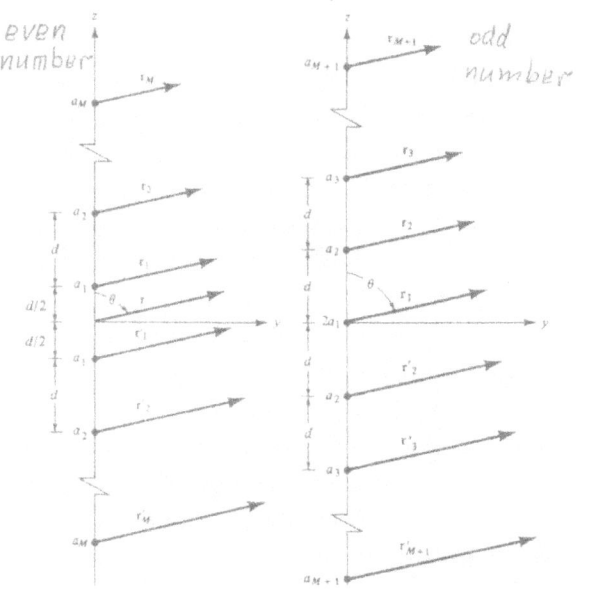

fig. 6.17 pp.291, Balanis

Examples of AFs of arrays of non-uniform amplitude distribution:

a) uniform amplitude distribution ($N=5$, $d = \lambda/2$, $\theta_0 = 90°$)

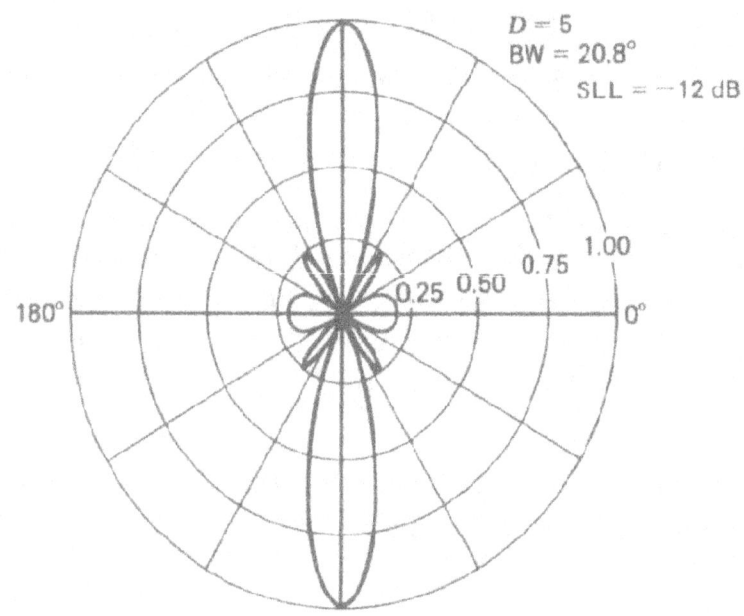

b) triangular (1:2:3:2:1) amplitude distribution ($N=5$, $d = \lambda/2$, $\theta_0 = 90°$)

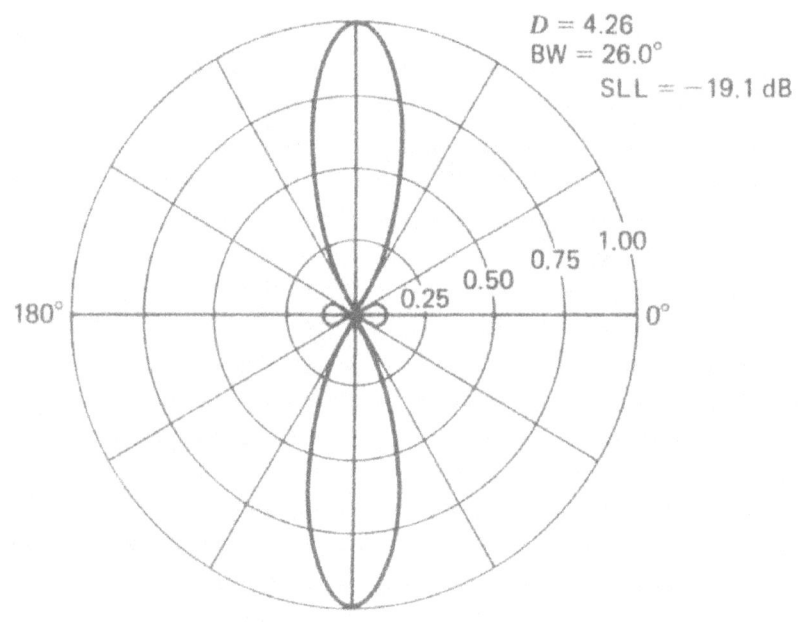

c) binomial (1:4:6:4:1) amplitude distribution ($N=5$, $d = \lambda/2$, $\theta_0 = 90°$)

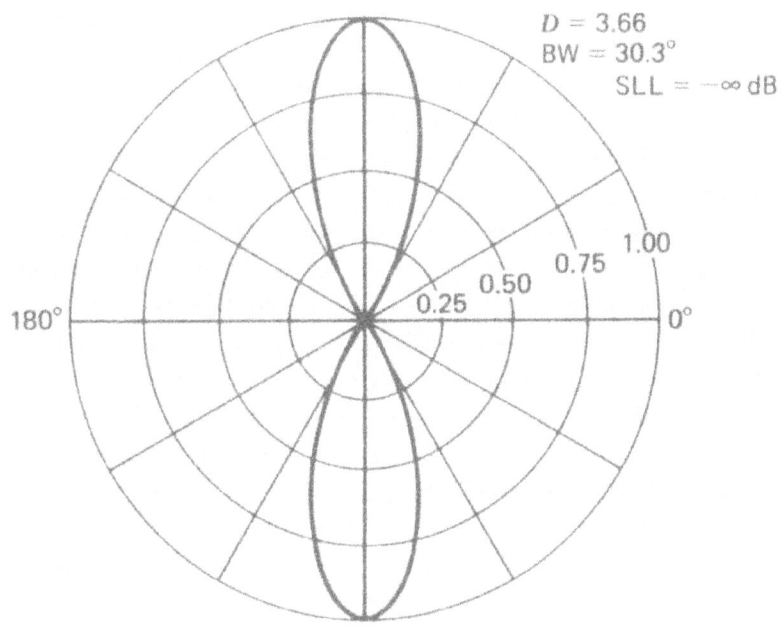

d) Dolph-Tschebyschev (1:1.61:1.94:1.61:1) amplitude distribution ($N=5$, $d = \lambda/2$, $\theta_0 = 90°$)

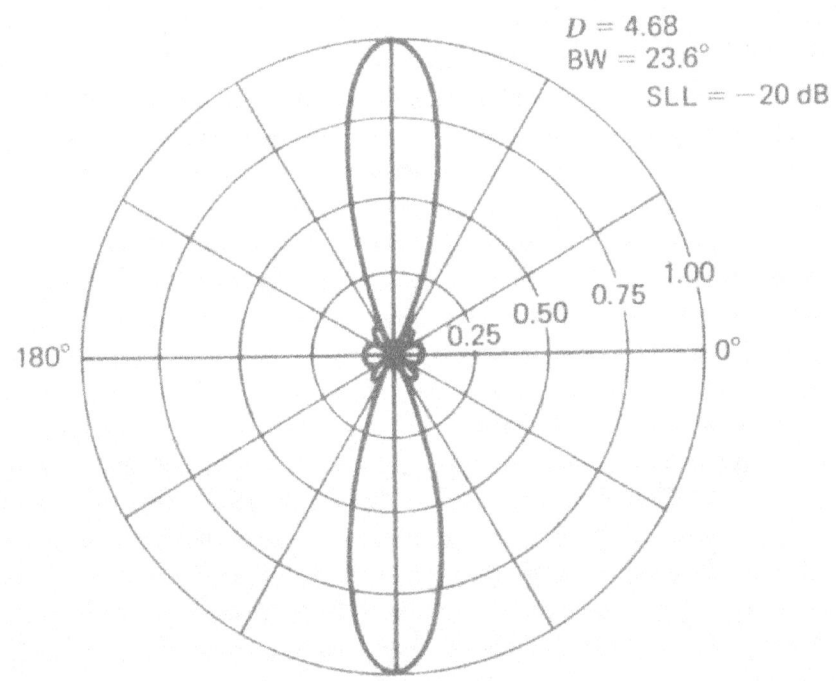

e) Dolph-Tschebyschev (1:2.41:3.14:2.41:1) amplitude distribution ($N=5$, $d = \lambda/2$, $\theta_0 = 90°$)

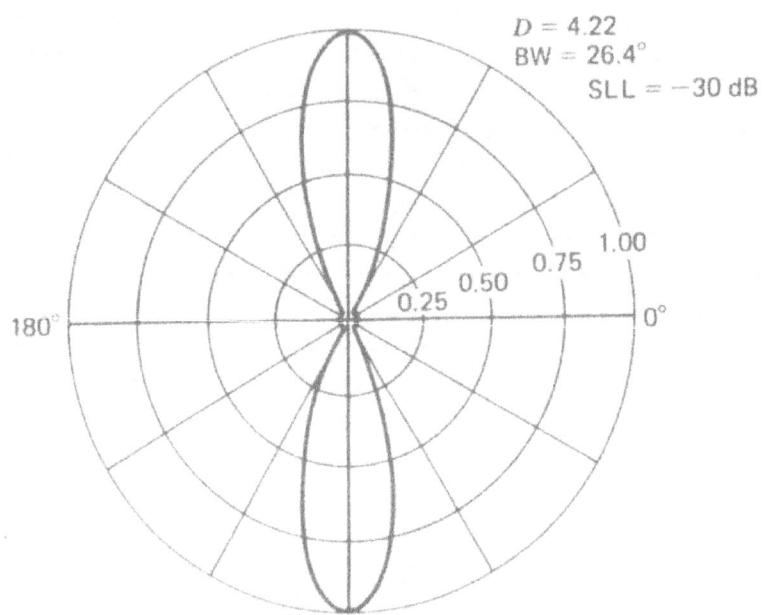

Notice that as the current amplitude is tapered more towards the edges of the array, the side-lobes tend to decrease, and the beamwidth tends to increase.

Let us consider a linear array with an even number ($2M$) of elements, located symmetrically along the z-axis, with excitation, which is also symmetrical with respect to $z = 0$. For a broadside array, ($\beta = 0$):

$$AF^e = a_1 e^{j\frac{1}{2}kd\cos\theta} + a_2 e^{j\frac{3}{2}kd\cos\theta} + ... + a_M e^{j\frac{2M-1}{2}kd\cos\theta} + \\ + a_1 e^{-j\frac{1}{2}kd\cos\theta} + a_2 e^{-j\frac{3}{2}kd\cos\theta} + ... + a_M e^{-j\frac{2M-1}{2}kd\cos\theta} \quad (17.1)$$

$$\Rightarrow AF^e = 2\sum_{n=1}^{M} a_n \cos\left[\left(\frac{2n-1}{2}\right)kd\cos\theta\right] \quad (17.2)$$

If the linear array consists of an odd number (2M+1) of elements, located symmetrically along the z-axis, then the array factor is:

$$AF^o = 2a_1 + a_2 e^{jkd\cos\theta} + a_3 e^{j2kd\cos\theta} + \ldots + a_{M+1} e^{jMkd\cos\theta} +$$
$$+ a_2 e^{-jkd\cos\theta} + a_3 e^{-j2kd\cos\theta} + \ldots + a_{M+1} e^{-jMkd\cos\theta} \quad (17.3)$$

$$\Rightarrow AF^o = 2\sum_{n=1}^{M+1} a_n \cos\left[(n-1)kd\cos\theta\right] \quad (17.4)$$

The factors (2) in equations (17.2) and (17.4) are unimportant for the normalized AF. Equations (17.2) and (17.4) can be re-written as:

$$AF^e = \sum_{n=1}^{M} a_n \cos\left[(2n-1)u\right], \text{ where } N = 2M \quad (17.5)$$

$$AF^o = \sum_{n=1}^{M+1} a_n \cos\left[2(n-1)u\right], \text{ where } N = 2M+1 \quad (17.6)$$

Here, $u = \dfrac{\pi d}{\lambda}\cos\theta$.

3. Binomial array

The binomial BSA was investigated and proposed by J.S. Stones to synthesize patterns without side lobes. First, consider a 2–element array (along the z-axis).

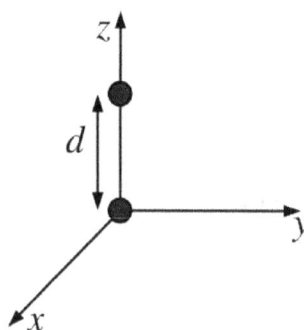

The elements of the array are identical and their excitation is the same. Its array factor is of the form:

$$AF = 1 + Z, \text{ where } Z = e^{j\psi} = e^{j(kd\cos\theta + \beta)} \quad (17.7)$$

If the spacing is $d \leq \lambda/2$ and $\beta = 0$ (broad-side maximum), this array will have no side lobes at all.

Second, consider a 2–element array whose elements are identical and the same as the array given above. The distance between the two arrays is again d.

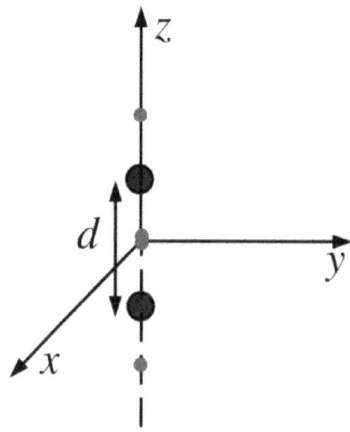

This new array has an AF of the form:
$$AF = (1+Z)(1+Z) = 1 + 2Z + Z^2 \qquad (17.8)$$
Since $(1+Z)$ has no side lobes, $(1+Z)^2$ will not have side lobes either.

Continuing the process for an N-element array produces:
$$AF = (1+Z)^{N-1} \qquad (17.9)$$
If $d \leq \lambda/2$, the above AF will not have side-lobes regardless of the number of elements N. The excitation amplitude distribution can be obtained easily by the expansion of the binome in (17.9). Making use of Pascal's triangle:

```
              1
            1   1
           1  2  1
          1  3  3  1
         1  4  6  4  1
        1  5 10 10  5  1
        ......................
```

the relative excitation amplitudes at each element of an (N+1)-element array can be determined. Such an array with a binomial distribution of the excitation amplitudes is called a *binomial array*. The current (excitation) distribution as given by the binomial expansion gives the *relative* values of the amplitudes. It is immediately seen that there is too wide variation of the amplitude, which is the major disadvantage of the BAs. The overall efficiency of such antenna would be very low. Besides, the BA has relatively wide beam. Its HPBW is the largest as compared to this of the uniform BSA or the Dolph–Chebyshev array.

Approximate closed-form expression for the HPBW of a BA with $d = \lambda/2$ is:

$$HPBW = \frac{1.06}{\sqrt{N-1}} = \frac{1.06}{\sqrt{2L/\lambda}} = \frac{1.75}{\sqrt{L/\lambda}} \quad (17.10)$$

where $L = (N-1)d$ is the array's length. The AFs of 10-element broadside binomial arrays ($N=10$) are given below.

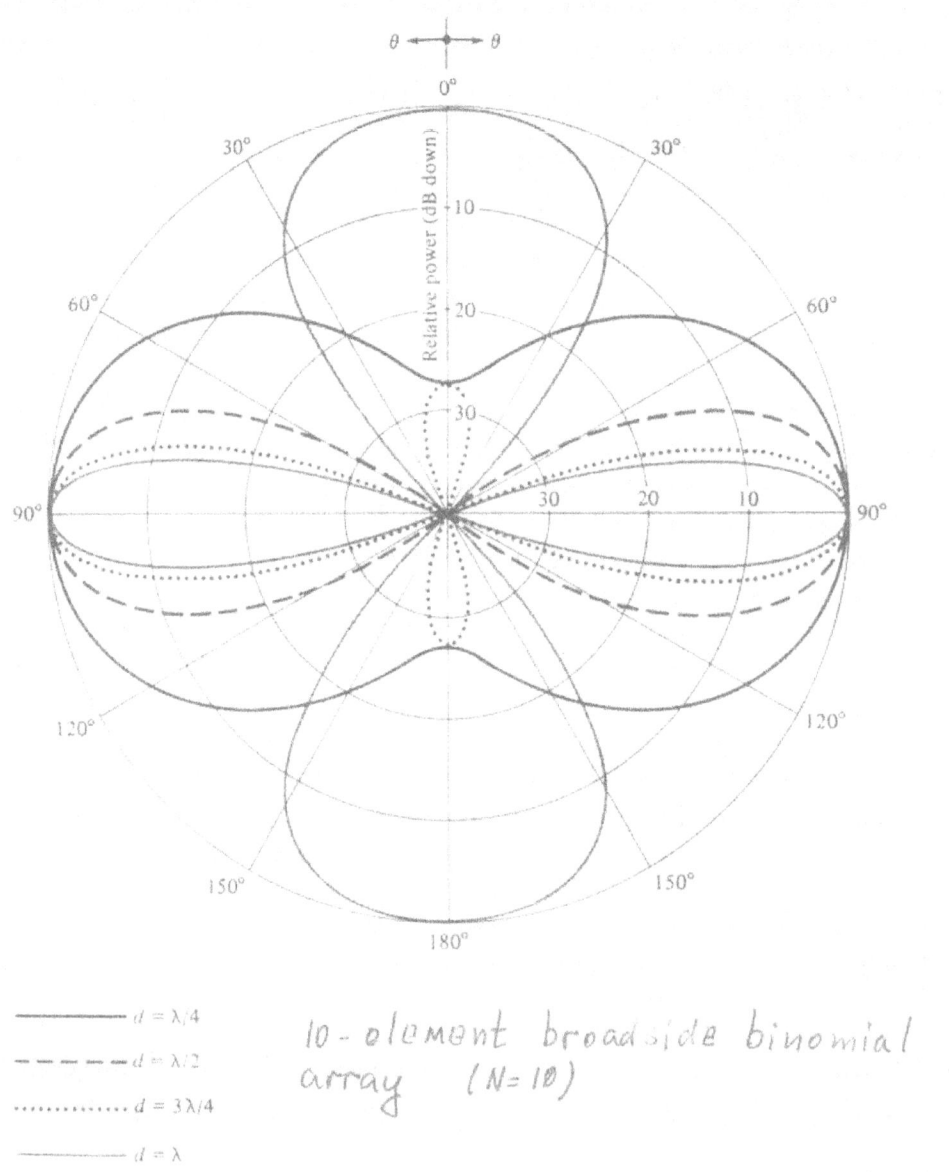

———— $d = \lambda/4$
— — — $d = \lambda/2$
············ $d = 3\lambda/4$
———— $d = \lambda$

10-element broadside binomial array (N=10)

Fig. 6.18, pp.293, Balanis

The directivity of a broadside BA with spacing $d = \lambda/2$ can be calculated from the formula below:

$$D_0 = \frac{2}{\int_0^\pi \left[\cos\left(\frac{\pi}{2}\cos\theta\right)\right]^{2(N-1)} d\theta}\bigg|_{d=\lambda/2} \quad (17.11)$$

$$D_0 = \frac{(2N-2)(2N-4)...2}{(2N-3)(2N-5)...1} \quad (17.12)$$

$$D_0 = 1.77\sqrt{N} = 1.77\sqrt{1 + 2L/\lambda} \quad (17.13)$$

4. Dolph–Chebyshev (DCA)

Chebyshev≡Tschebyscheff

Dolph proposed (in 1946) a method to design arrays with any desired side-lobe levels and any HPBWs. This method is based on the approximation of the pattern of the array by a Chebyshev polynomial of order m, high enough to meet the requirement for the side-lobe levels. Actually, a DCA with no side lobes (side-lobe level of $-\infty$ dB) reduces to the binomial design.

4.1 The Chebyshev polynomials

The Chebyshev polynomials are defined by:

$$T_m(z) = \begin{cases} (-1)^m \cosh(m\cosh^{-1}|z|), & z < -1 \\ \cos(m\cos^{-1} z), & -1 < z < 1 \\ \cosh(m\cosh^{-1} z), & z > 1 \end{cases} \quad (17.14)$$

A nice feature of Chebyschev polynomials is that $T_m(z)$ of any order m can be derived via a recursion formula, provided $T_{m-1}(z)$ and $T_{m-2}(z)$ are defined.

$$T_m(z) = 2zT_{m-1}(z) - T_{m-2}(z) \qquad (17.15)$$

Explicitly, (17.15) produces:

$$\begin{aligned} m &= 0, \; T_0(z) = 1 \\ m &= 1, \; T_1(z) = z \\ m &= 2, \; T_2(z) = 2z^2 - 1 \\ m &= 3, \; T_3(z) = 4z^3 - 3z \\ m &= 4, \; T_4(z) = 8z^4 - 8z^2 + 1 \\ m &= 5, \; T_5(z) = 16z^5 - 20z^3 + 5z, \; \text{etc.} \end{aligned} \qquad (17.16)$$

If $|z| < 1$, then Chebyshev polynomials are related to the cosine functions, see (17.14). One can always expand the function $\cos(mx)$ as a polynomial of $\cos(x)$ of order m, e.g.,

$$\cos 2x = 2\cos^2 x - 1 \qquad (17.17)$$

This is done by making use of Euler's formula:

$$(e^{jx})^m = (\cos x + j\sin x)^m = e^{jmx} = \cos(mx) + j\sin(mx) \qquad (17.18)$$

Similar relations hold for the hyperbolic cosine function. From the example (17.17), one can see that the Chebyshev argument z is related to the cosine argument x by:

$$z = \cos x \quad \text{or} \quad x = \arccos z \qquad (17.19)$$

Then (17.17) can be written as:

$$\cos(2\arccos z) = 2[\cos(2\arccos z)]^2 - 1$$
$$\Rightarrow \cos(2\arccos z) = 2z^2 - 1 = T_2(z) \qquad (17.20)$$

Compare it with definition (17.14) or with (17.16)-line 3.

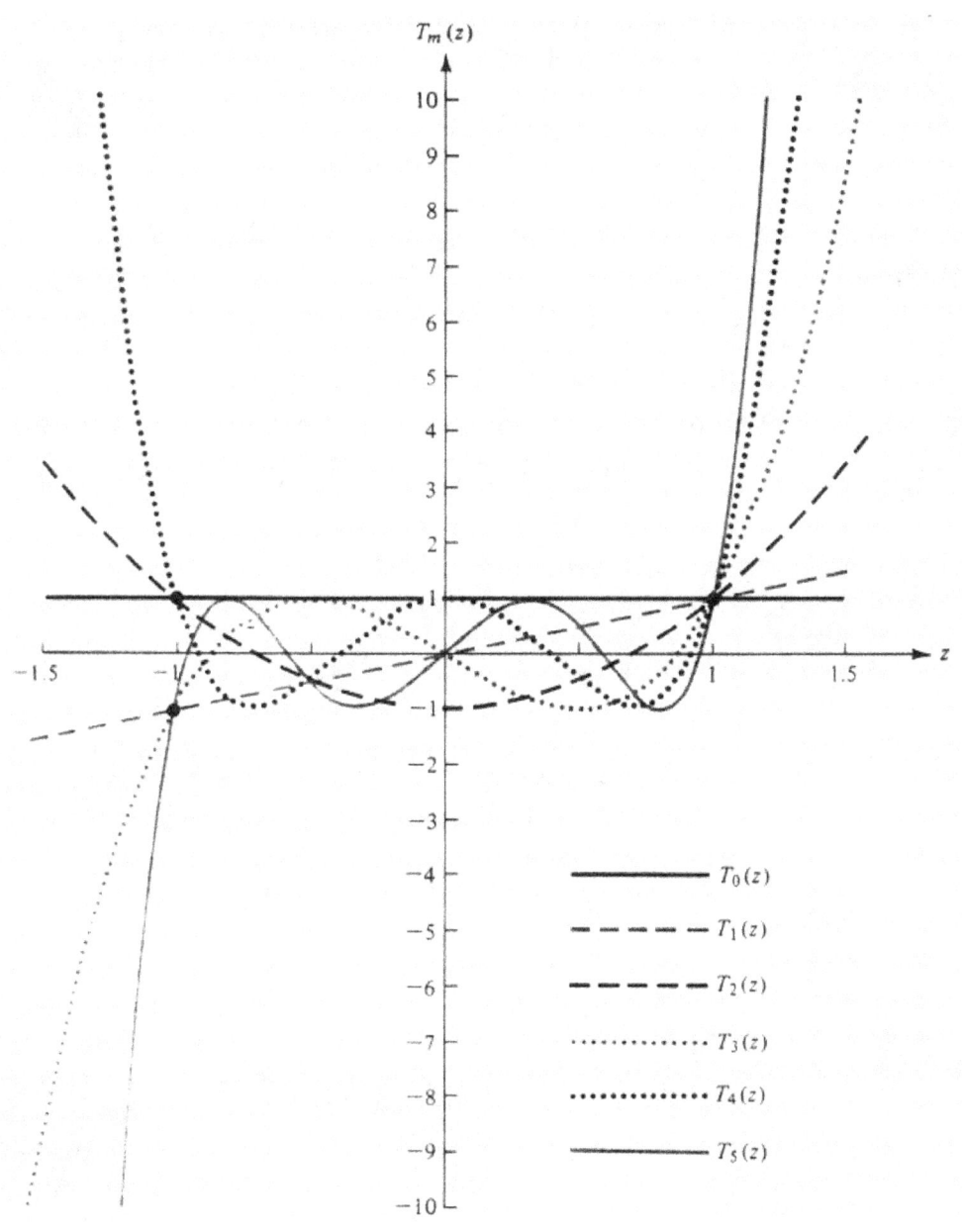

Fig. 6.19, pp.296, Balanis

Properties of Chebyshev polynomials:
1) All polynomials of any order m pass through the point (1,1).
2) Within the range $-1 \leq z \leq 1$, the polynomials have values within [-1,1].
3) All nulls occur within $-1 \leq z \leq 1$.
4) The maxima and minima in the $z \in [-1,1]$ range have values +1 and –1, respectively.
5) The higher the order of the polynomial, the steeper the slope for $|z|>1$

4.2 Chebyshev array design
The main goal is to approximate the desired AF with a Chebyshev polynomial such that
- the side-lobe level meets the requirements, and
- the main beam width is small enough.

An array of N elements has an AF, which can be approximated with a Chebyshev polynomial of order m that is always:

$$\boxed{m = N - 1}, \quad (17.21)$$

where: $N = 2M$, if N is even;
$N = 2M + 1$, if N is odd.

In general, for a given side-lobe level, the higher the order of the polynomial, the narrower the beamwidth. But for $m>10$, the difference is not substantial (see the slopes of $T_m(z)$ in the previous figure).

The AF of an N-element array (17.5) or (17.6) will be identical with a Chebyshev polynomial if:

$$T_{N-1}(z) = \begin{cases} \sum_{n=1}^{M} a_n \cos[(2n-1)u], & N = 2M \text{ -even} \\ \sum_{n=1}^{M+1} a_n \cos[2(n-1)u], & N = 2M+1 \text{-odd} \end{cases} \quad (17.22)$$

Here, $u = \frac{\pi d}{\lambda} \cos\theta$.

Let the side-lobe level be:

$$R_0 = \frac{E_{\max}}{E_{sl}} = \frac{1}{AF_{sl}} \text{ (voltage ratio)} \quad (17.23)$$

Then the maximum of T_{N-1} is fixed at an argument z_0, where

$$T_{N-1}^{\max}(z_0) = R_0, \quad (17.24)$$

where $T_{N-1} > 1$.

Equation (17.24) corresponds to $AF(u) = AF^{\max}(u_0)$. Obviously, z_0 must satisfy the condition:

$$z_0 > 1 \quad (17.25)$$

Then, the portion of $AF(u)$, which corresponds to $T_{N-1}(z)$ for $|z| < 1$, will have levels lower or equal to the specified side-lobe level R_0. This portion of AF must correspond to the out-ot-main-beam radiation pattern, i.e. the side lobes. The AF is a polynomial of $\cos u$ and the $T_{N-1}(z)$ is a polynomial of z where the limits for z are:

$$-1 \leq z \leq z_0 \quad (17.26)$$

Since
$$-1 \leq \cos u \leq 1 \quad (17.27)$$
the relation between z and $\cos u$ must be set as:
$$\cos u = \frac{z}{z_0} \quad (17.28)$$
where $z_0 > 1$.

Array design for an array of N elements – general procedure
1) Expand the AF as given by (17.5) or (17.6) by replacing each $\cos(mu)$ term ($m = 1, 2, ..., M$) with the power series of $\cos u$.
2) Determine z_0 such that $T_{N-1}(z_0) = R_0$ (voltage ratio).
3) Substitute $\cos u = z/z_0$ in the AF as found in the previous step.
4) Equate the AF found in Step 3 to $T_{N-1}(z)$ and determine the coefficients for each power of z.

Example: Design a DCA (broadside) of $N=10$ elements with a major-to-minor lobe ratio of $R_0 = 26$ dB. Find the excitation coefficients and form the AF.

Solution:
1. The AF is:
$$AF_{2M} = \sum_{n=1}^{5} a_n \cos[(2n-1)u], \quad u = \frac{\pi d}{\lambda} \cos\theta$$
2. Expand AF_{2M} in terms of $\cos u$:
$$AF_{10} = a_1 \cos u + a_2 \cos 3u + a_3 \cos 5u + a_4 \cos 7u + a_5 \cos 9u$$
Here:
$$\cos 3u = 4\cos^3 u - 3\cos u$$
$$\cos 5u = 16\cos^5 u - 20\cos^3 u + 5\cos u$$
$$\cos 7u = 64\cos^7 u - 112\cos^5 u + 56\cos^3 u - 7\cos u$$
$$\cos 9u = 256\cos^9 u - 576\cos^7 u + 432\cos^5 u - 120\cos^3 u + 9\cos u$$

3. Determine z_0:

$$R_0 = 26 \text{ dB} \Rightarrow R_0 = 10^{26/20} \simeq 20$$
$$\Rightarrow T_9(z_0) = 20$$
$$\cosh\left[9\cosh^{-1}(z_0)\right] = 20$$
$$9\cosh^{-1}(z_0) = \cosh^{-1} 20 = 3.69$$
$$\cosh^{-1}(z_0) = 0.41$$
$$z_0 = \cosh 0.41$$
$$z_0 = 1.08515$$

4. Express the AF in terms of $z = z_0 \cos u$:

$$AF_{10} = \frac{z}{z_0}(a_1 - 3a_2 + 5a_3 - 7a_4 + 9a_5)$$

$$+ \frac{z^3}{z_0^3}(4a_2 - 20a_3 + 56a_4 - 120a_5)$$

$$+ \frac{z^5}{z_0^5}(160a_3 - 112a_4 + 432a_5)$$

$$+ \frac{z^7}{z_0^7}(64a_4 - 576a_5)$$

$$+ \frac{z^9}{z_0^9}(256a_5) = \underbrace{9z - 120z^3 + 432z^5 - 576z^7 + 256z^9}_{T_9(z)}$$

5. Finding the coefficients by matching the power terms:

$$256a_5 = 256z_0^9 \Rightarrow a_5 = 2.0860$$
$$64a_4 - 576a_5 = -576z_0^7 \Rightarrow a_4 = 2.8308$$
$$16a_3 - 112a_4 + 432a_5 = 432z_0^5 \Rightarrow a_3 = 4.1184$$
$$4a_2 - 20a_3 + 56a_4 - 120a_5 = -120z_0^7 \Rightarrow a_2 = 5.2073$$

$$a_1 - 3a_2 + 5a_3 - 7a_4 + 9a_5 = 9z_0^9 \Rightarrow a_1 = 5.8377$$

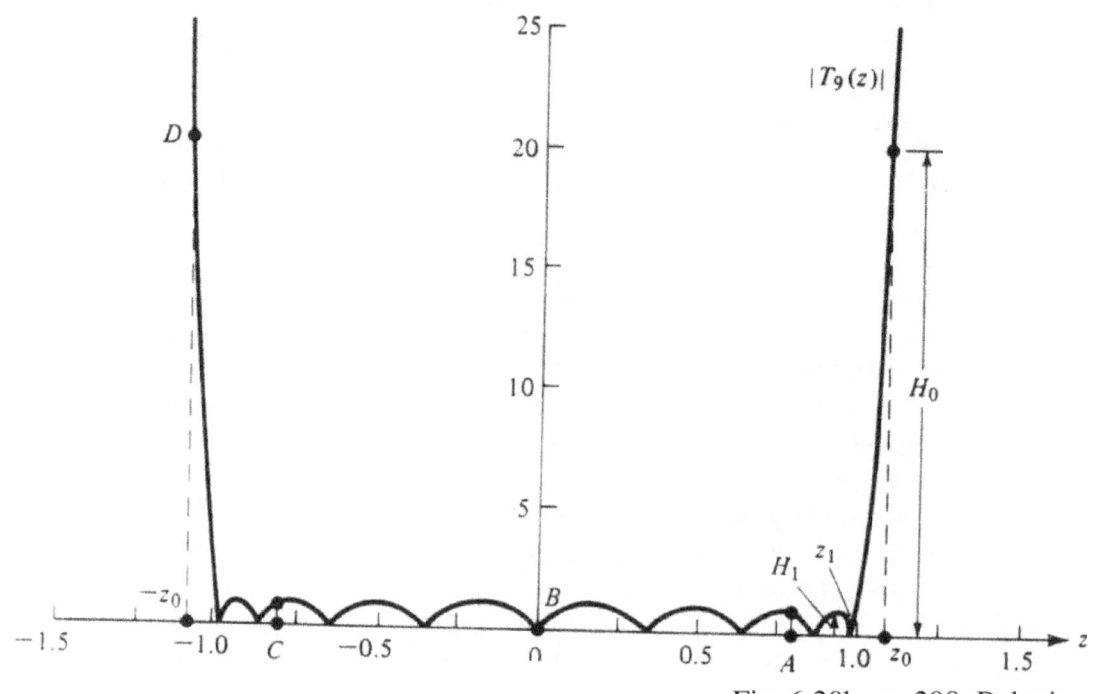

Fig. 6.20b, pp.298, Balanis

6. Normalize coefficients with respect to edge element ($N=5$):
   $a_5 = 1$; $a_4 = 1.357$; $a_3 = 1.974$; $a_2 = 2.496$; $a_1 = 2.789$

$$AF_{10} = 2.789\cos(u) + 2.496\cos(3u) + 1.974\cos(5u)$$
$$+ 1.357\cos(7u) + \cos(9u)$$

where $u = \dfrac{\pi d}{\lambda}\cos\theta$.

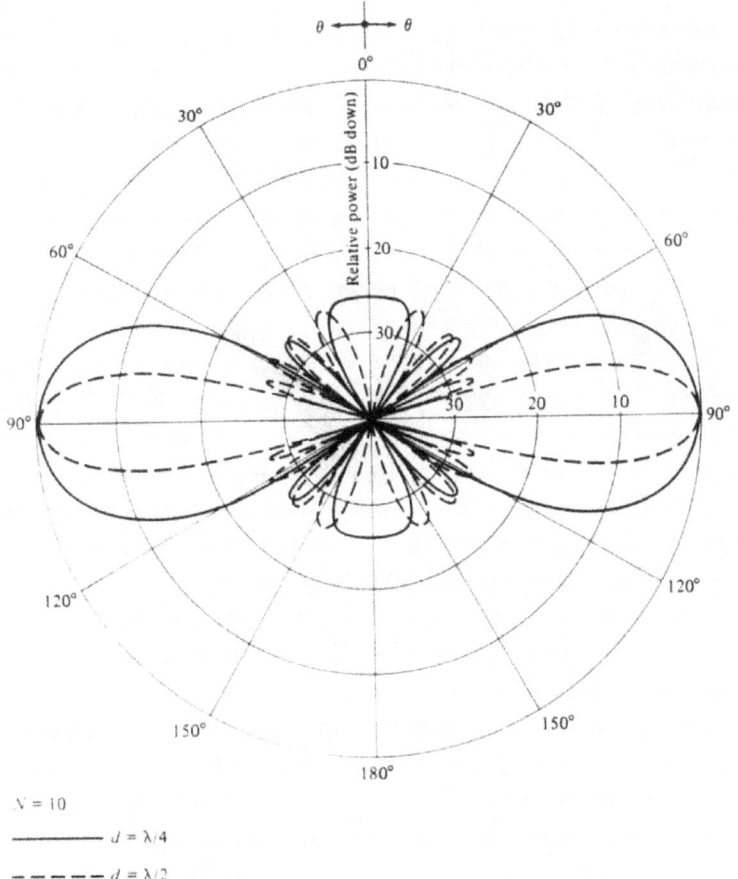

Fig. 6.21, pp.300, Balanis

4.3. The maximum affordable d, d_{max}, for Chebyshev arrays.

This restriction arises from the requirement for a single major lobe – see also equation (17.26).

$$z \geq -1$$
$$\Rightarrow z_0 \cos\left(\frac{\pi d}{\lambda}\cos\theta\right) \geq -1 \qquad (17.29)$$

For a given array, when θ varies from $0°$ to $180°$, the argument z assumes values:

$$\text{from } z = z_0 \cos\frac{\pi d}{\lambda} \text{ to } z = z_0 \cos\left(-\frac{\pi d}{\lambda}\right)$$

The extreme value z to the left of the abscissa is $z = z_0 \cos\frac{\pi d}{\lambda}$ (end-fire directions of the AF, $\theta = 0°$ or $180°$). This value must not go beyond $z = -1$; otherwise minor lobes of levels higher than 1 (higher than R_0) will appear. Therefore, the inequality (17.29) must hold for $\theta = 0°$ or $180°$:

$$z_0 \cos\left(\frac{\pi d_{max}}{\lambda}\right) \geq -1 \Rightarrow \cos\left(\frac{\pi d_{max}}{\lambda}\right) \geq -\frac{1}{z_0}$$

Let:

$$\gamma = \arccos\left(\frac{1}{z_0}\right)$$

Then:

$$\frac{\pi d_{max}}{\lambda} < \pi - \gamma = \pi - \arccos\left(\frac{1}{z_0}\right)$$

$$\Rightarrow \boxed{\frac{d_{max}}{\lambda} < 1 - \frac{1}{\pi}\arccos\left(\frac{1}{z_0}\right)} \quad (17.30)$$

In the previous example:

$$\frac{d_{max}}{\lambda} < 1 - \frac{1}{\pi}\arccos\left(\frac{1}{1.08515}\right) = 1 - \frac{0.39879}{\pi} = 0.873$$

$$\underline{d_{max} < 0.873\lambda}$$

5. Directivity of non-uniform arrays

It is difficult to derive closed form expressions for the directivity of non-uniform arrays. Here, we shall derive expressions in the form of series in the most general case of a linear array.

The unnormalized array factor is:

$$AF = \sum_{n=0}^{N-1} a_n e^{j\beta_n} e^{jkz_n \cos\theta} \quad (17.31)$$

where:

a_n is the amplitude of the excitation of the *n*-th element;

β_n is the phase angle of the excitation of the *n*-th element;

z_n is the z-coordinate of the *n*-th element.

The maximum *AF* is:

$$AF_{max} = \sum_{n=1}^{N-1} a_n \qquad (17.32)$$

The normalized *AF* is:

$$AF_n = \frac{AF}{AF_{max}} = \frac{\sum_{n=0}^{N-1} a_n e^{j\beta_n} e^{jkz_n \cos\theta}}{\sum_{n=1}^{N-1} a_n} \qquad (17.33)$$

The beam solid angle is derived as:

$$\Omega_A = 2\pi \int_0^\pi |AF_n(\theta)|^2 \sin\theta \, d\theta$$

$$\Omega_A = \frac{2\pi}{\left(\sum_{n=1}^{N-1} a_n\right)^2} \sum_{m=0}^{N-1}\sum_{p=0}^{N-1} a_m a_p e^{j(\beta_m-\beta_p)} \int_0^\pi e^{jk(z_m-z_p)\cos\theta} \sin\theta \, d\theta$$

where:

$$\int_0^\pi e^{jk(z_m-z_p)\cos\theta} \sin\theta \, d\theta = \frac{2\sin\left[k(z_m-z_p)\right]}{k(z_m-z_p)}$$

$$\Omega_A = \frac{4\pi}{\left(\sum_{n=1}^{N-1} a_n\right)^2} \sum_{m=0}^{N-1}\sum_{p=0}^{N-1} a_m a_p e^{j(\beta_m-\beta_p)} \frac{\sin\left[k(z_m-z_p)\right]}{k(z_m-z_p)} \qquad (17.34)$$

$$D_0 = \frac{4\pi}{\Omega_A}$$

$$\Rightarrow D_0 = \frac{\left(\sum_{n=1}^{N-1} a_n\right)^2}{\sum_{m=0}^{N-1}\sum_{p=0}^{N-1} a_m a_p e^{j(\beta_m-\beta_p)} \frac{\sin[k(z_m-z_p)]}{k(z_m-z_p)}} \quad (17.35)$$

For equispaced LA (17.35) reduces to:

$$D_0 = \frac{\left(\sum_{n=1}^{N-1} a_n\right)^2}{\sum_{m=0}^{N-1}\sum_{p=0}^{N-1} a_m a_p e^{j(\beta_m-\beta_p)} \frac{2\sin[(m-p)kd]}{(m-p)kd}} \quad (17.36)$$

because $z_n = nd$.

For equispaced broadside arrays, where $\beta_m = \beta_p$ for any (m,p), equation (17.36) reduces to:

$$D_0 = \frac{\left(\sum_{n=1}^{N-1} a_n\right)^2}{\sum_{m=0}^{N-1}\sum_{p=0}^{N-1} a_m a_p \frac{\sin[(m-p)kd]}{(m-p)kd}} \quad (17.37)$$

For equispaced broadside uniform arrays:

$$D_0 = \frac{N^2}{\sum_{m=0}^{N-1}\sum_{p=0}^{N-1} \frac{2\sin[(m-p)kd]}{(m-p)kd}} \quad (17.38)$$

When the spacing d is a multiple of $\lambda/2$, equation (17.37) reduces to:

$$D_0 = \frac{\left(\sum_{n=1}^{N-1} a_n\right)^2}{\sum_{n=0}^{N-1} (a_n)^2}, \quad d = \frac{\lambda}{2}, \lambda, \ldots \qquad (17.39)$$

Example: Calculate the directivity of the Dolph–Chebyshev array designed in the previous example if $d = \lambda/2$.

The 10-element DCA has the following amplitude distribution:
$$a_5 = 1; \quad a_4 = 1.357; \quad a_3 = 1.974; \quad a_2 = 2.496; \quad a_1 = 2.798$$
We make use of (17.39):

$$D_0 = \frac{4\left(\sum_{n=1}^{5} a_n\right)^2}{2\sum_{n=0}^{2} (a_n)^2} = 2\frac{(9.625)^2}{20.797} = 8.9 \quad (9.5 \text{ dB}) \qquad (17.40)$$

6. Half-power beamwidth of a BS DCA.

For large DCA with side lobes in the range (-20 to –60) dB, the HPBW can be found by introducing a beam-broadening factor, f, given by:

$$f = 1 + 0.636 \left\{ \frac{2}{R_0} \cosh\left[\sqrt{(\text{arccos} R_0)^2 - \pi^2} \right] \right\}^2 \qquad (17.41)$$

The HPBW of the DCA is equal to the product of the broadening factor by the HPBW of the respective uniform linear array:

$$HPBW_{DCA} = f \times HPBW_{UA} \qquad (17.42)$$

In (17.41) R_0 denotes the side-lobe level (voltage ratio).

Two Slopers for All Traditional HF- Bands

By Vladimir Fursenko, UA6CA

It is possible at one mast install two slopers that cover all traditional five HF-Bands- 80,- 40,- 40,- 15 and 10- meter. **Figure 1** shows the antenna.

One sloper covers 80 and 10- meter Bands another one covers 40, 20 and 15- meter Bands. Each sloper feed through separate coaxial cable. It is possible use 50 or 75-Ohm coaxial cable. Coaxial cable matched with antenna with help of length of two wire line. It may be open line or two wire line in plastic insulation.

For example, at the **Figure 1** it is shown at right- two wire open line and at left- two wire ribbon line. Line in plastic insulation should be shortened (factor 0.82) compare to two wire open line.

Coaxial cable may have any length. However, coaxial cable in length 13.65 or 27.30- meter for 40, 20 and 15- meter Band antenna and coaxial cable in length 27.0- meter for 80 and 10- meter Band antenna would give good result. The antenna gives over 0.5- 1.0- dB compare to VS1AA.

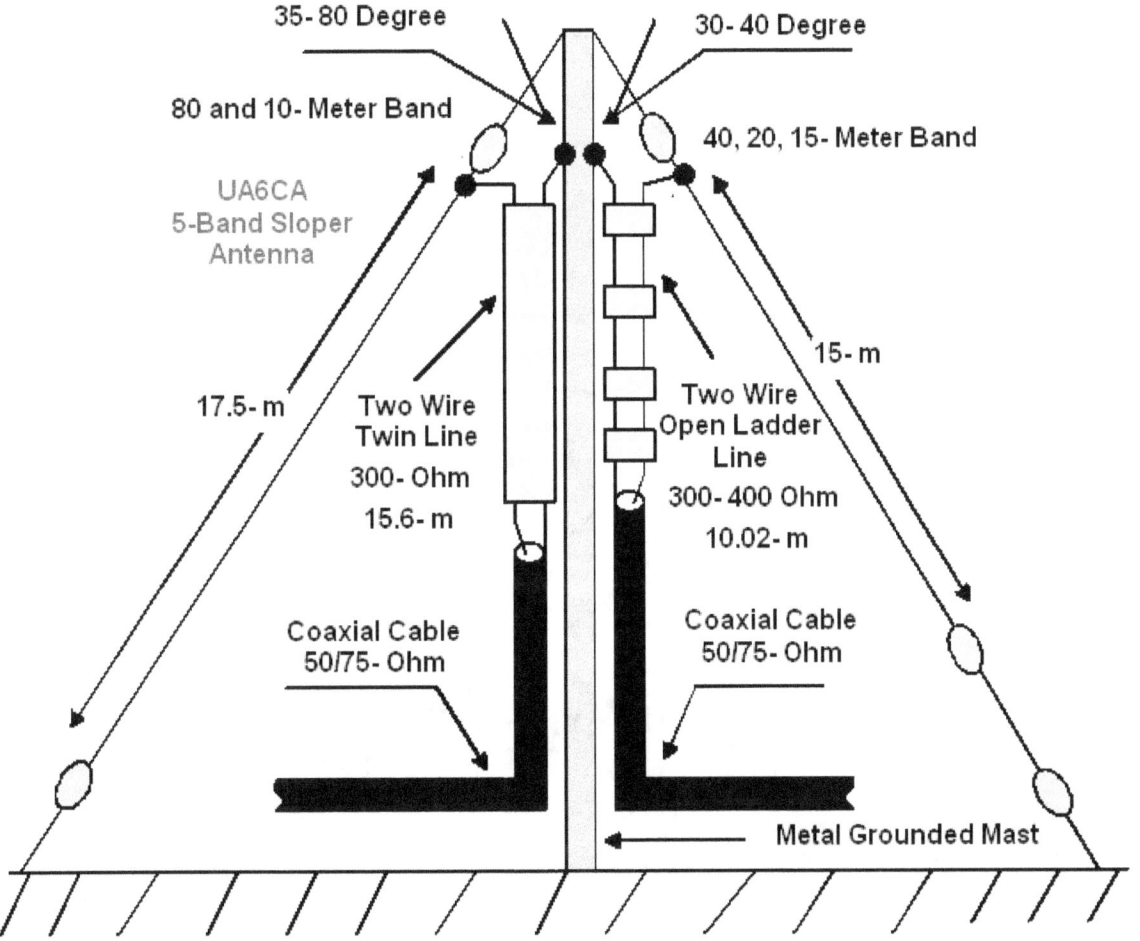

Figure 1 Two Slopers for All Traditional Five HF-Bands

Antenna for 80-, 40,- 20,- 17,- 15,- 12 and 10- meter Band

By Vladimir Fursenko, UA6CA

The two- level antenna works at 80-, 40-, 20-, 17-, 15-, 12- and 10- meter Bands. Upper level works on 80-, 40-, 20-, 15- and 10- meter Bands. Lower level works on 17- and 12- meter Bands. **Figure 1** shows design of the antenna.

Upper level has input impedance of 180… 220- Ohm. Lower level has input impedance of 50-Ohm. Each level feeds by its own coaxial cable. **Figure 2** shows feeding of the levels. Antenna at the upper level is fed through a transformer with ratio 1:4.

The transformer is wound on a ferrite ring with OD 60- mm and height 10-mm. The ring may have permeability 400… 600. Transformer has 10 turns wound by a pair of wire in diameter 1.5- mm (15- AWG). Then the turns are connected with antenna and together accordingly to **Figure 2**. The antenna may be fed through one coaxial cable. It needs to install RF Relay at the lower level that would be turn the cable to a chosen antenna. The RF Relay should switch both- central core and braid of the coaxial cable.

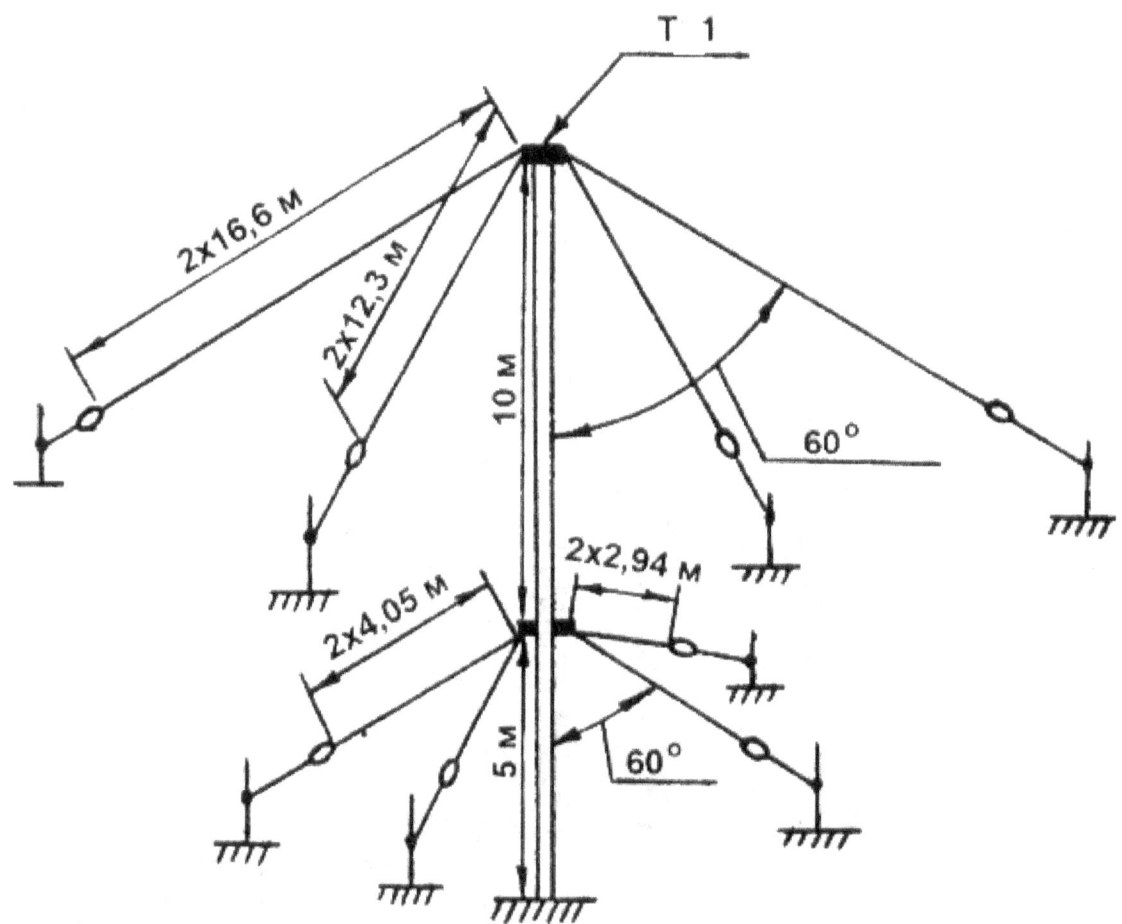

Figure 1 Design of the Antenna for 80-, 40-, 20-, 17-, 15-, 12-, and 10- meter HF Band

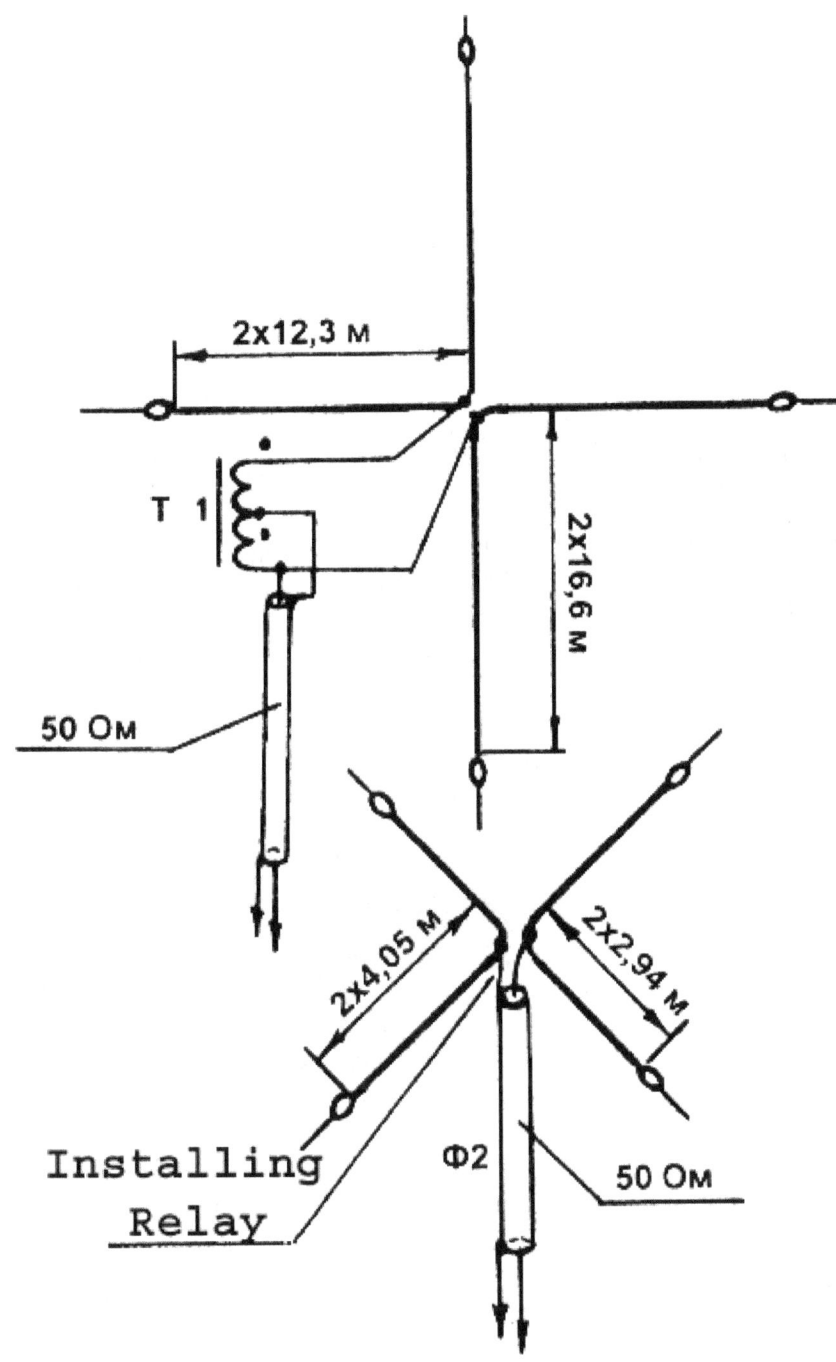

Figure 2 Feeding of each Level of the Antenna for 80-, 40-, 20-, 17-, 15-, 12-, and 10- meter HF Band

Bidirectional Vertical Antenna for the 20- meter Band

The publication is devoted to the memory UR0GT.

Credit Line: Forum from:
www.cqham.ru

By: Nikolay Kudryavchenko, UR0GT

One Ground Plane allows get one directional switchable DD. For this purpose to the GP a two wires should be added. With help an RF Relay (or just with help of manually installed jumper) the wires turn on to Director and Reflector. Length of the GP should be taken a little more than lambda/4. For the minimum SWR the antenna tune up with the help of a capacitor C1.

The MMANA model of the Bidirectional Vertical Antenna for the 20- meter Band may be loaded: http://www.antentop.org/019/ur0gt_directional_019.htm

Figure 1 shows the design of the antenna.

Figure 2 shows Z of the Bidirectional Vertical Antenna for the 20- meter Band.

Figure 3 shows SWR of the Bidirectional Vertical Antenna for the 20- meter Band.

Figure 4 shows DD of the Bidirectional Vertical Antenna for the 20- meter Band.

73! de UR0GT

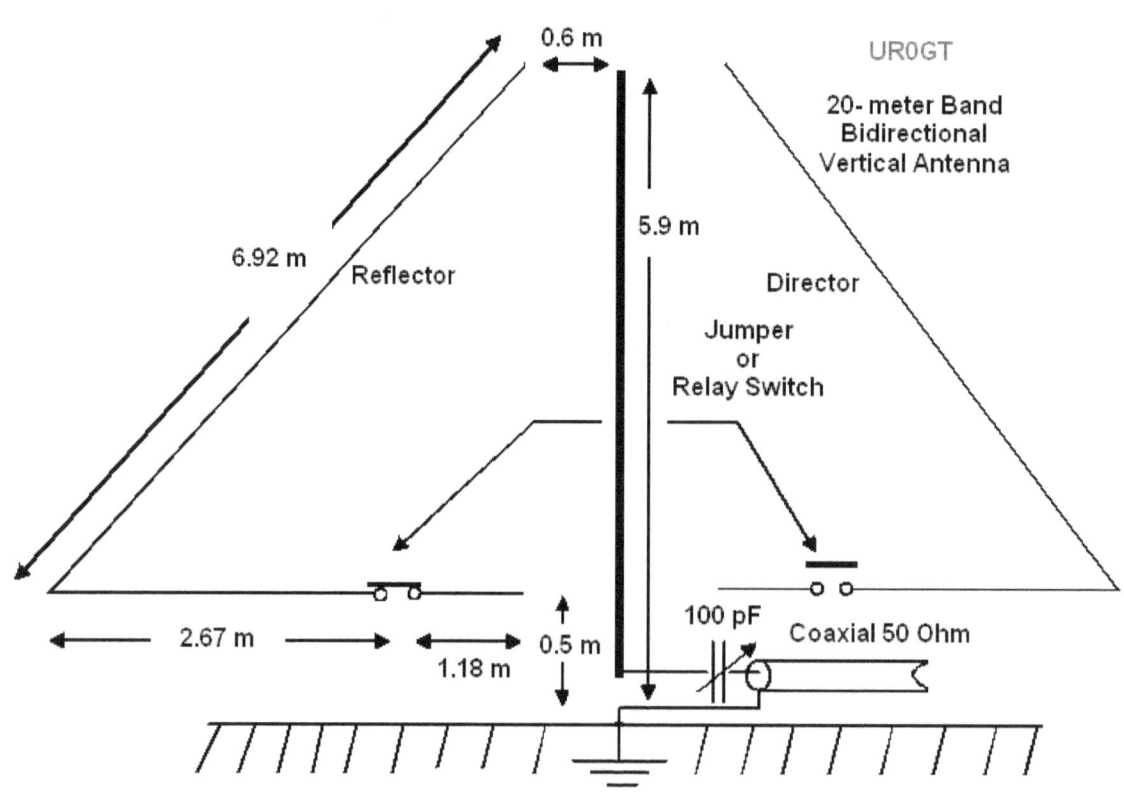

Figure 1 Design of the Bidirectional Vertical Antenna for the 20- meter Band

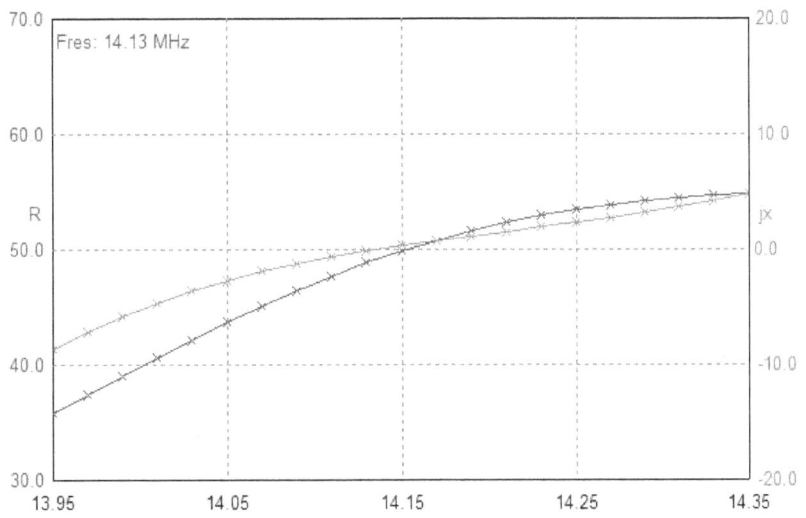

Figure 2 Z of the Bidirectional Vertical Antenna for the 20- meter Band

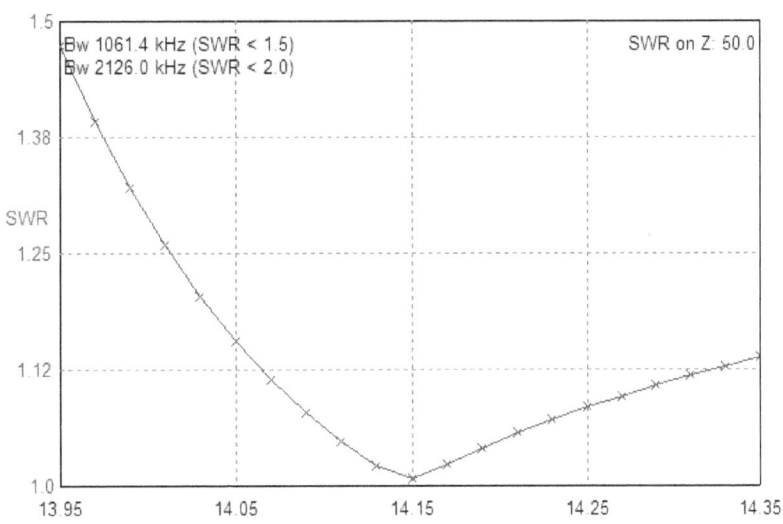

Figure 3 SWR of the Bidirectional Vertical Antenna for the 20- meter Band

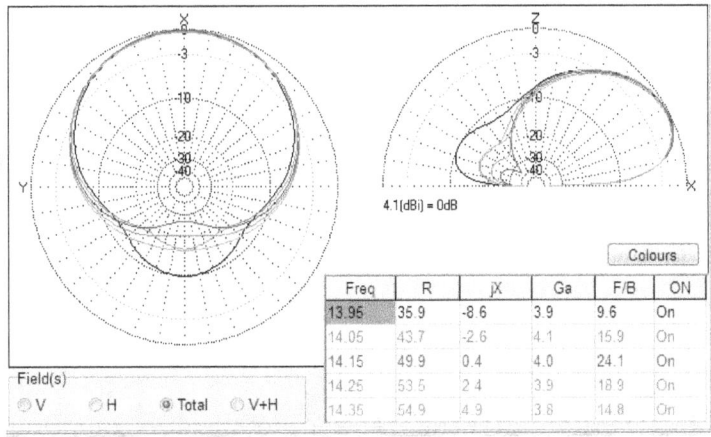

Figure 4 DD of the Bidirectional Vertical Antenna for the 20- meter Band

L- Vertical Antenna for nearest objects for the 40- and 20- meter Band

The publication is devoted to the memory UR0GT.

Credit Line: Forum from:
www.cqham.ru

By: Nikolay Kudryavchenko, UR0GT

One L-Vertical Antenna may works at the two amateur Bands- 40- and 20- meters. The main feature of the antenna is that it may be located at nearest conductive objects- for example the antenna may be placed near water drain. Antenna may be placed at a wall of a house. Inside a house there are always a lot of conductive objects- for example electrical main, refrigerator, metal tubes of house ventilation system and so on.

Parts L1 and L2 of the antenna intended for tuning the antenna into resonance at the 20- meter Band. Length of the parts may be changed at real antenna installation. For the antenna a good ground system is required. However as usual it is not a big problem to make a ground system beside such antenna. For example, the antenna was simulated with wire counterpoises in 5 meter length at height 0.1- meter however in real those ones may be placed straight away to the ground.

The MMANA model of the L-Vertical Antenna for Nearest Objects for The 40- and 20- meter Bands may be loaded: http: // www.antentop.org/019/ur0gt_L_Vertical_019.htm

Figure 1 Shows design of the L-Vertical Antenna.

Figure 2 Z of the L-Vertical Antenna for the 40- meter Band.

Figure 3 Shows SWR of the L-Vertical Antenna for the 40- meter Band.

Figure 4 Shows DD of the L-Vertical Antenna for the 40- meter Band

Figure 5 Shows Z of the L-Vertical Antenna for the 20- meter Band.

Figure 6 Shows SWR of the L-Vertical Antenna for the 20- meter Band.

Figure 7 Shows DD of the L-Vertical Antenna for the 20- meter Band.

73! de UR0GT

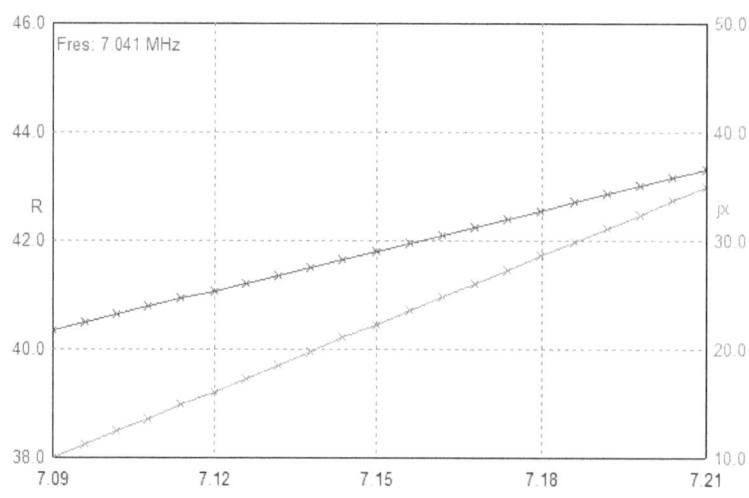

Figure 2 Z of the L-Vertical Antenna for the 40- meter Band

Figure 1 Design of the L-Vertical Antenna

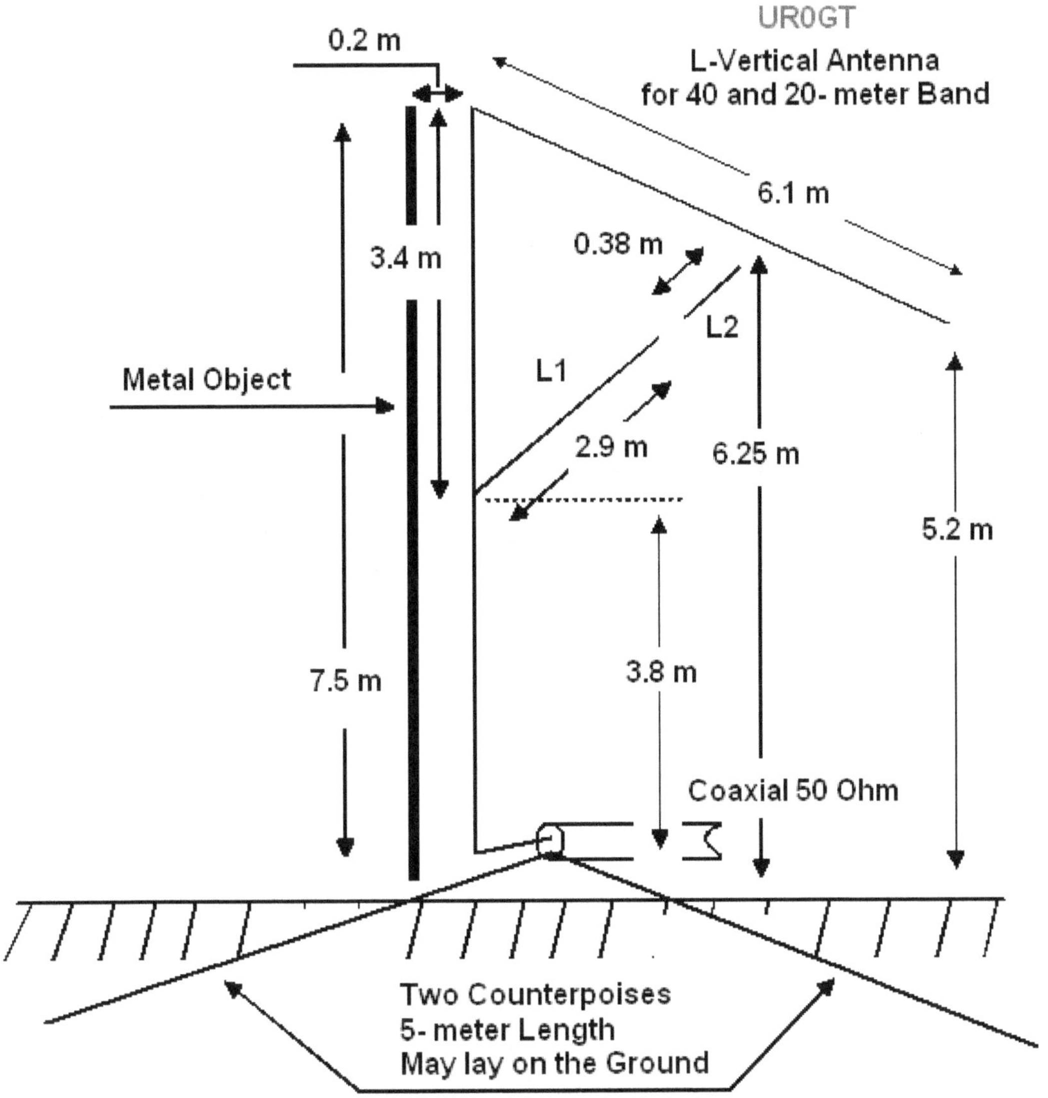

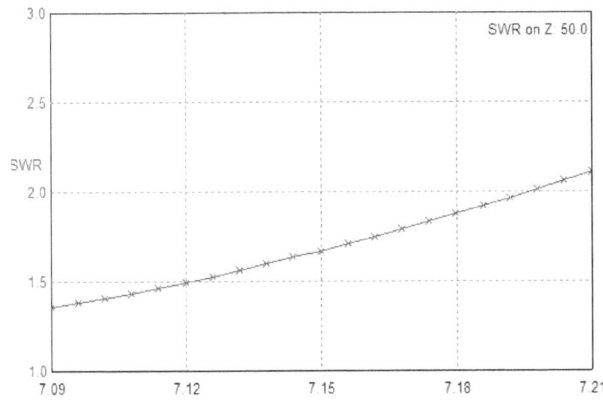

Figure 3 SWR of the L-Vertical Antenna for the 40-meter Band

Freq	R	jX	Ga	F/B	ON
7.09	40.4	10.0	0.7	-1.8	On
7.12	41.1	16.2	0.7	-1.6	On
7.15	41.8	22.3	0.7	-1.8	On
7.18	42.6	28.7	0.7	-1.8	On
7.21	43.3	35.0	0.7	-1.8	On

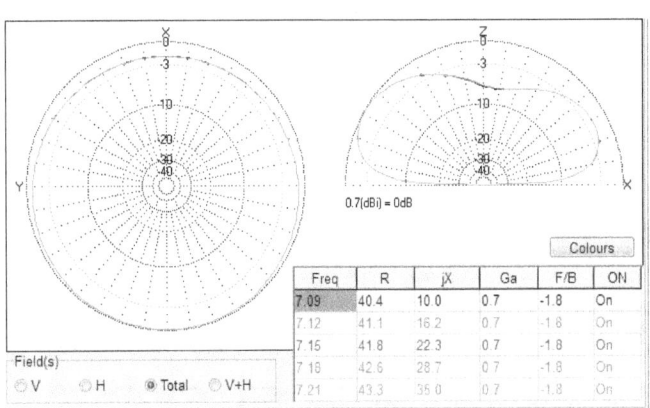

Figure 4 DD of the L-Vertical Antenna for the 40-meter Band

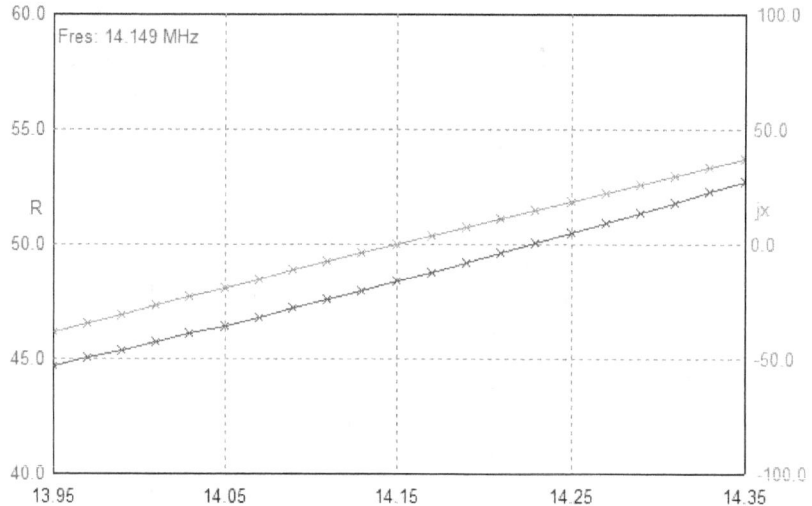
Figure 5 Z of the L-Vertical Antenna for the 20- meter Band

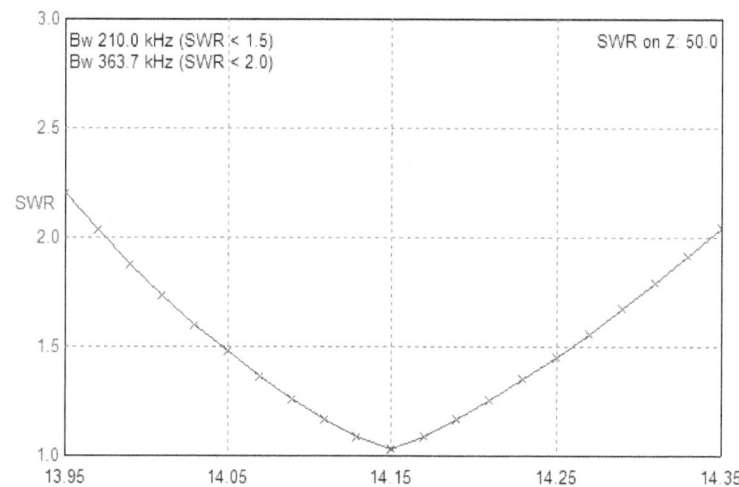
Figure 6 SWR of the L-Vertical Antenna for the 20- meter Band

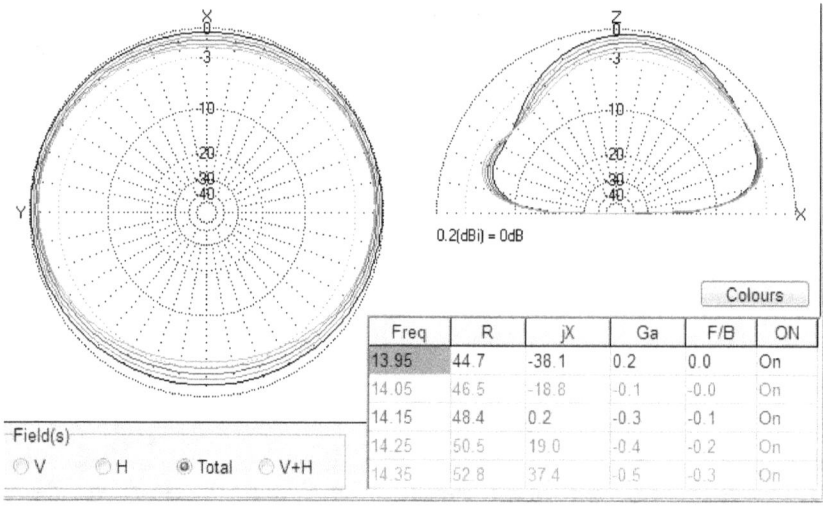
Figure 7 DD of the L-Vertical Antenna for the 20- meter Band

Universal Beverage Antenna

Igor Grigorov, va3znw, Richmond Hill, Canada

Beverage Antennas are widely used at commercial and military radio communication. In commercial communication Beverage Antenna as usual is used as a receiving antenna. However, in military communication Beverage Antenna is used for both purposes- for receiving and transmitting applications. Transmitting/receiving Beverage Antenna was used in DX- Pedition EK1NWB on to Kizhy island (the antenna described at: http://www.antentop.org/008/ua3znw008.htm) where the antenna (against skepticism of some persons) illuminated its good job.

So when again in Toronto I have changed my QTH and the QTH allowed me install Beverage Antenna, I did not hesitated.

Beverage Antenna has lots advantages that attractive me. *First*, it is low noise receiving antenna. At all my previously settled QTHs I had so devastated noise level that 160 and 80 meter Bands were closed for me. *Second*, Beverage Antenna is lighting safety antenna because the antenna wire grounded from both sides and the antenna wire is placed at small height above the ground. *Third*, Beverage Antenna is sustained at strong winds and ice rain- it is very important for Canadian winter.

Forth, Beverage Antenna is (at proper installation) practically invisible. That is very important in the place where some antennas may be restricted. *Fifth*, Beverage Antenna is very broadband antenna. Without any ATU the antenna may have good SWR on all amateurs HF Bands from 160 to 6 meter. *Sixth*, Beverage Antenna has single lobe diagram directivity. It is possible count again and again the advantages of the Beverage Antenna.

But we begin count disadvantages. *First and the main* lack of the antenna is the low efficiency on to transmission. However, the lack may be easy improved with PA- but if you do not hear anything (usual matter in modern city overloaded by electromagnetic smog) you do not need PA...

Figure 1 shows a Classical Beverage Antenna. Beverage Antenna consists of a horizontal wire with length L. The length may be from one-half to tens wavelengths long. The wire suspended above the ground at height H. For real receiving antennas the height may be from 1.5 up to 5.0 meter. For military transmitting Beverage Antennas the height may be from 0.5 up to 1.5 meter.

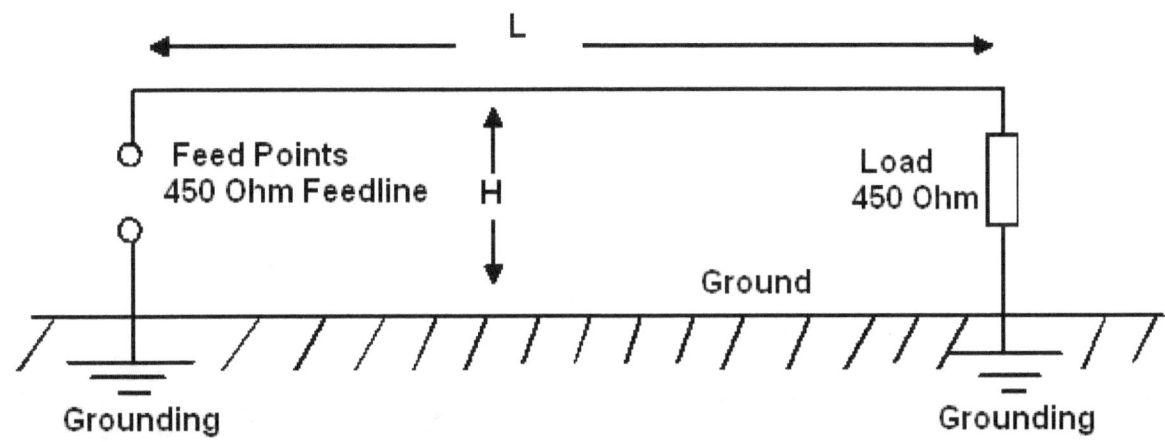

L: Antenna Length. Several Wavelength

H: Antenna Height. 1.5 ... 2- meter above the Ground

Figure 1 Classical Beverage Antenna

Receiver/transmitter with the feedline is attached to one end and the other terminated through a resistor (300- 600 Ohm) to ground. The value of the resistor should be equal to the wave impedance of the transmission line that created by the antenna wire and the ground under the wire. So, optimum value of the resistor depends on the height the wire above the ground and diameter of the wire. Beverage Antenna has SWR close to 1.0:1.0 in a wide spectrum of the radio frequency band when the optimum value resistor is used and the antenna fed through line with impedance that is equal to the wave impedance of the antenna.

Beverage Antenna has practically unidirectional radiation pattern (small back lobe of course is present) with the main lobe directed to the resistor-terminated end. The wide of the lobe depends on to ratio "used frequency: antenna length." I believe that the paragraph gave enough theoretical knowledge to build a Beverage Antenna.

Figure 2 shows Beverage Antenna that I have installed at my backyard. Length of my Beverage antenna was equal to 20- meters. The length was fixed by the length of the fence. Antenna wire was stapled to the fence. I used to a 16- AWG (1.3- mm) wire in strong black plastic insulation. The wire was bought at sale at Home Depot- 22 cent/meter. The wire used to in electrical job. The antenna wire was placed at height 1.8- meter above the ground. Height of the fence defines it. Theoretical value of the wave impedance of my antenna was close to 500- Ohm. It is allowed me use 450- Ohm terminated resistor and transformer 1:9 to feed the antenna through coaxial cable RG8X 50- Ohm. At my case the cable had length in 50 feet. At both end sides the Beverage Antenna had RF and electrical grounding.

Below we discuss all parts of the Beverage Antenna.

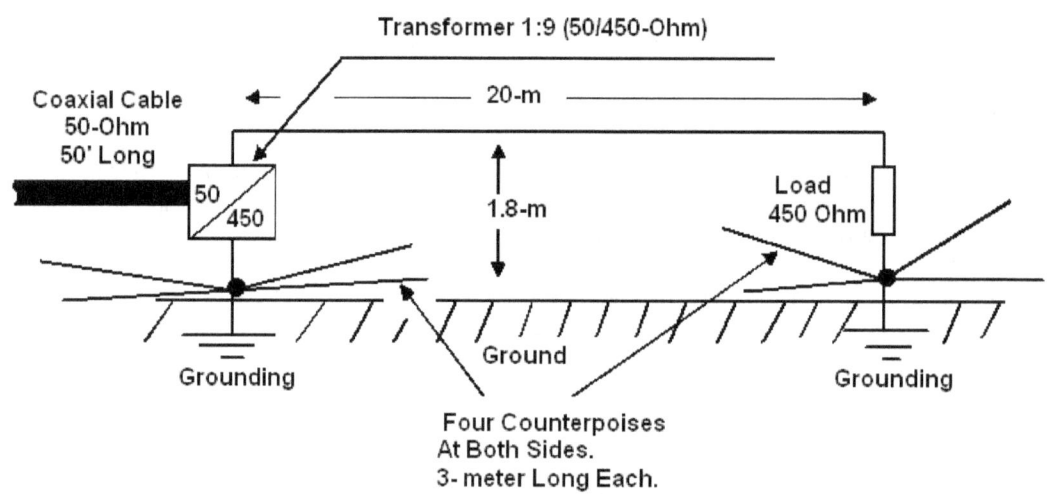

Figure 2 Practical Design of my Beverage Antenna

Terminated resistor of the Beverage Antenna

I need termination resistor for my Beverage Antenna. It should be 450- Ohm, non-inductive, 20… 50- Wtts Broadband termination resistor. Power of the resistor depends on output power of used transmitter. The resistor may dissipate up to 50% of the RF power going to the antenna. It is not a problem to buy such resistor online through internet. However, the chipper one costs $ 50.0 USD. I decided make the termination resistor by myself. I bought a kit with 25 e.a. usual 4.7-kOm/2.5-Wtt metal resistors for $ 5.0 on e-bay. Eleven such resistors switched to bridge have resistance 440- Ohm. **Figure 3** shows the home- brew termination resistor. Coin in 25- cents is placed beside the termination resistor. Dissipative power of the resistor should be 27.5- Wtts. However my experiments show that the resistor could stand at least 50- Wtts for a short time.

So, the Beverage antenna may work with transmitter with output power 100- Wtts with CW and SSB mode.

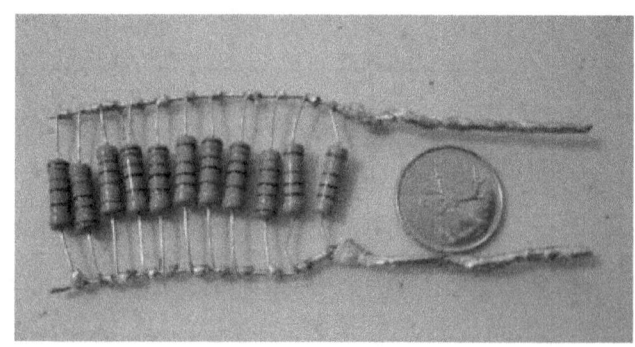

Figure 2 Practical Design of Home- Brew Termination Resistor

Transformer of the Beverage Antenna

Transformer of the Beverage Antenna is one of the important parts of the antenna. Transformer should work at all frequency range of the antenna. Transformer should stand power going to the antenna. It is preferably use to transformer with insulated windings for any Beverage Antenna. **Figure 4** shows Beverage Antenna with transformer with insulated windings. Such transformer provides electrical insulation the antenna from transceiver. It causes less noise from electrical interferences and provides additional protection of the transceiver from lightning discharge. However to make such transformer is not an easy task. Thereof I used unsymmetrical RF transformer 1:9.

Figure 5 shows schematic of classical unsymmetrical RF transformer 1:9. The transformer is winding by a triple wire onto a ferrite core. It may be a ferrite ring or ferrite rod. Transformer may contain 7-... 15 turns.

Quantity of turns depends onto size and permeability of the ferrite core and frequency range of the antenna.

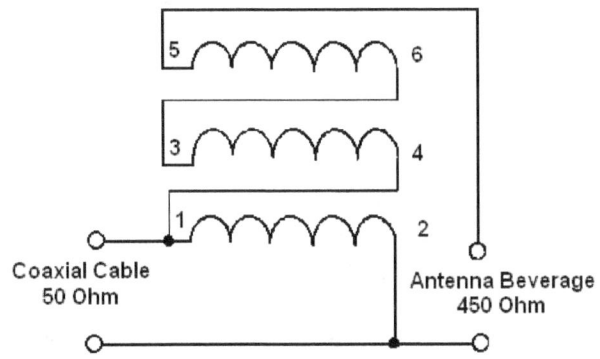

Figure 5 Schematic of Classical Unsymmetrical RF Transformer 1:9

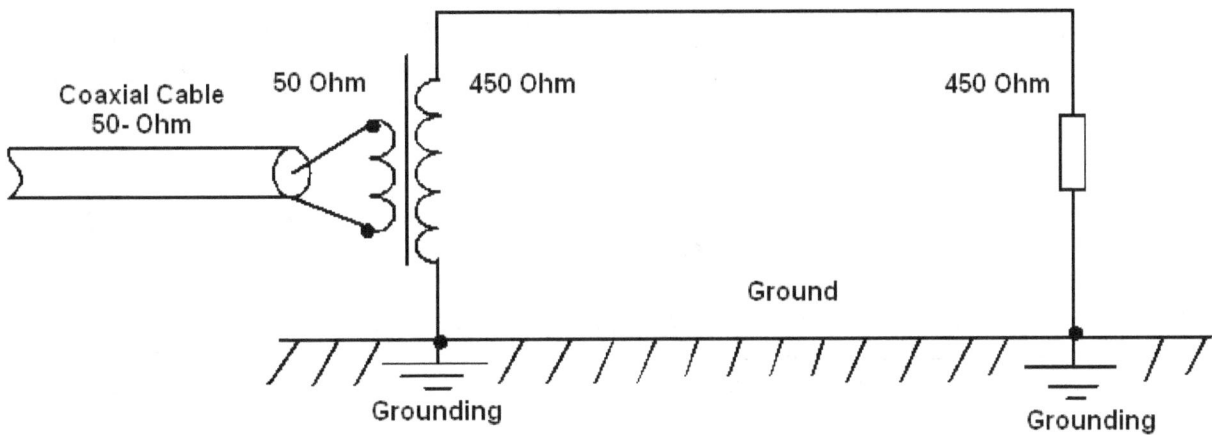

Figure 4 Beverage Antenna with Transformer with Insulated Windings

My first attempt to make RF transformer was use a TV Yoke's ferrite ring for the transformer 1:9. **Figure 6** shows the TV Yoke. I removed wires from the yoke. Then the ring was wrapped by black vinyl insulation tape. Unsymmetrical RF Transformer 1:9 was made according to the **Figure 5**. Transformer contained 8- turns of usual electrical wire. **Figure 7** shows ready Unsymmetrical RF Transformer 1:9 on TV Yoke. The transformer was tested. At the beginning the transformer was tested with usual metal resistor in resistance 440 Ohm. **Table 1** shows data for transformer made on Yoke Core loaded to Resistor 440 Ohm the measurements made by MFJ- 259B. **Table 2** shows data for the transformer made on Yoke Core loaded to home brew Resistor 440 Ohm (shown at **Figure 3**) the measurements made by MFJ- 259B. The home brew resistor has some reactance that clearly seen from the two tables. MFJ- 259B does not indicates the character of the reactance (capacitance or inductive) so I used sign "@" at the reactance.

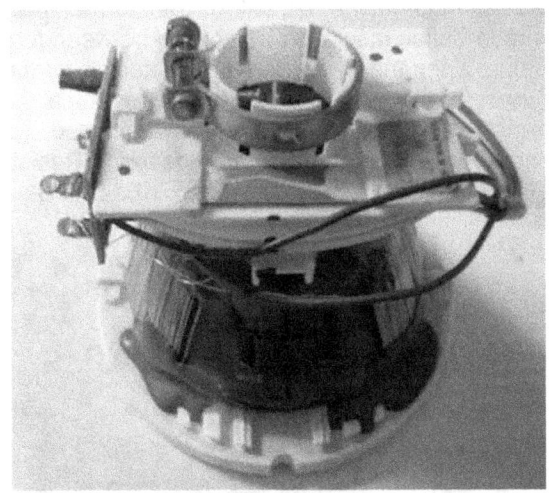

Figure 6 TV Yoke

Table 1 and Table 2 show that the transformer works well at 160- and 80- meter Bands, fair at 40- and 20- meter Bands and bad beginning from 17- meter Band. Of course it is possible play with quantity of the turns but anyway it should be difficult to make a real broadband transformer that covers 160- 10- meter Bands. So I gave up attempt to use RF transformer on TV Yoke core at my Beverage Antenna.

In my parts box there was a symmetrical Balun from LDG. **Figure 8** shows the Balun. Some time ago I experimented with the Balun. However for that time it just took place in the box waiting for a new application. Time is come. The Balun had strong plastic cabinet with socket SO- 239 on it from one side and two terminals (for antenna and for ground) at another side.

Figure 7 Unsymmetrical RF Transformer 1:9 on TV Yoke

Table 1
Transformer made on Yoke Core. Loaded on to single resistor 440 Ohm. Measurement by MFJ- 259B

Band	160	80	40	30	20	17	15	12	10
Z	48@j5	45@j4	35@j1.4	27@j1	19@j11	13@j29	11@j40	9@j54	6@j69
SWR	1.1	1.1	1.4	1.8	2.5	4.3	6.7	9.1	12.4

Table 2
Transformer made on Yoke Core. Loaded on to eleven 4.7-kOm resistors switched to bridge (440 Ohm overall). Measurement by MFJ- 259B

Band	160	80	40	30	20	17	15	12	10
Z	49@j4	44@j6	31@j3	23@j2	15@j17	10@j33	10@j41	6@j61	5@j70
SWR	1.0	1.1	1.5	2.0	3.0	5.7	7.9	11.7	14.4

A ferrite ring was used on the Balun. The sizes of the ring were suite for transformer 1:9 that could work at 100- Wtts RF power going through. It was not hard reworked the Balun to unsymmetrical RF transformer 1:9. Additional wire in Teflon insulation was coiled between turns of the Balun. Then the wires were connected according to **Figure 5**. **Figure 9** shows the unsymmetrical RF transformer 1:9. The transformer was tested. **Table 3** shows data for the transformer loaded to Resistor 440 Ohm the measurements made by MFJ- 259B.

Table 4 shows data for the transformer loaded to home brew Resistor 440 Ohm (shown at **Figure 3**) the measurements made by MFJ- 259B. As you can see the transformer works well from 160- to 10- meter band. It is possible play with quantity of the turns to move low SWR up or down inside the 160- 10- meter Band. However, I did not do it and leaved things like this.

Figure 8 Balun RBA- 1:1 from LDG

Figure 9 Unsymmetrical RF transformer 1:9 on the Base of Balun RBA- 1:1

Table 3

Transformer made on LDG Balun Core. Loaded on to single resistor 440 Ohm. Measurement by MFJ- 259B

Band	160	80	40	30	20	17	15	12	10
Z	42@j17	48@j8	45@j3	40@j3	35@j3	31@j8	29@j12	29@j17	27@j29
SWR	1.5	1.1	1.1	1.1	1.4	1.6	1.8	2.0	2.4

Table 4

Transformer made on LDG Balun Core. Loaded on to eleven 4.7-kOm resistors switched to bridge (440 Ohm overall). Measurement by MFJ- 259B

Band	160	80	40	30	20	17	15	12	10
Z	43@j15	47@j7	42@j4	37Z@j4	31	27@j5	24@j10	23@j16	23@j21
SWR	1.4	1.1	1.2	1.3	1.6	1.8	2.1	2.5	2.8

Grounding of the Beverage Antenna

Right grounding of the each sides of Beverage Antenna is not a simple task. For example, at receiving centers of ex- USSR the grounding of Beverage Antenna was made by 10- 15 radial wires that were dig on to depth 20- 40- cm into the ground. It was good RF and electrical grounding. But it was not for me. I made simplified grounding. The grounding consisted of two parts (see **Figure 2**).

One part was *RF grounding*. The grounding made suction of RF currents from the antenna wire. There were four wires in length 3- meter each. I used the same wire that for Beverage Antenna. The wires were dig on to depth 5- cm into the ground.

Another part of the grounding there was *electrical grounding*. Electrical grounding allows static electricity flow from the antenna wire to the ground. As well the grounding increases the safety of the Beverage Antenna at lightning time. Antenna works quiet on to receiving with electrical grounding. **Figure 10** shows design of the electrical grounding. To make the grounding I used two copper strips (it was Copper Strapping from Home Depot) in length 3- feet and five copper tubes in length one foot each. The stuff was soldered according to the **Figure 10**.

I used compact Bernzomatic Butane Gas burner for soldering. **Figure 11** shows soldered grounding. **Figure 12** shows grounding at the antenna (feeding side). Ditch in depth near 5- cm was made in the place of the copper strip. The strips were placed into the ditches. Copper tubes were hammered into the ground. After that a batch of earth (with grass seed) was covered the electrical grounding. Grass already covers my grounding.

Design of the Beverage Antenna

Wire of the Beverage Antenna was stapled to the wooden fence. **Figure 13** shows stapled wire.

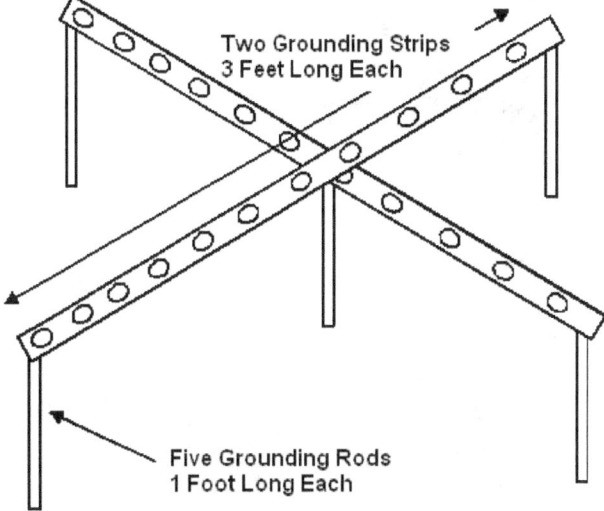

Figure 10 Design of the Electrical Grounding

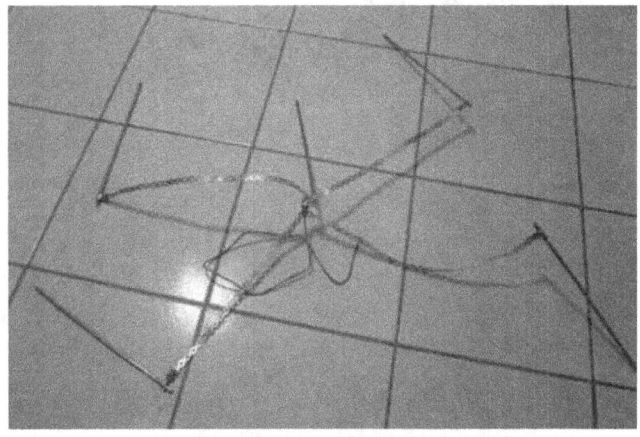

Figure 11 Soldered Grounding

Termination resistor and transformer of Beverage Antenna were placed into food plastic boxes.

The boxes were chosen to fit the resistor assembly and transformer. **Figure 14** shows termination resistor in open plastic food box. **Figure 15** shows transformer in open plastic food box. Each box was closed by cover.

Then several turns of a vinyl plastic tape (I used electrical tape for outdoor application that should work at temperature range: – 20 C to + 60 C) were coiled above the cover. **Figure 16** shows termination resistor in closed plastic food box. **Figure 17** shows transformer in closed plastic food box. Coaxial cable is going to basement through a window with plastic insert. The insert made from soft foam. **Figure 18** shows the plastic insert on the window. Near entry of the cable to the plastic insert the coaxial cable was coiled in a small coil. The coil was an RF Choke that closed way for stray RF currents inducted to outer jacket of the coaxial cable to the transceiver. Also the choke gave additional protection from lightning. **Figure 19** shows the RF choke.

Figure 14 Termination Resistor of Beverage Antenna in Open Plastic Food Box

Figure 12 Grounding at the Antenna

Figure 15 Transformer of Beverage Antenna in Open Plastic Food Box

Figure 13 Stapled Wire of Beverage Antenna

Figure 16 Termination Resistor of Beverage Antenna in Closed Plastic Food Box

Figure 17 Transformer of Beverage Antenna in Closed Plastic Food Box

Figure 18 Plastic Insert on the Window

Test of the Beverage Antenna

At first I have measured SWR of the Beverage Antenna with help of MFJ-259B and with internal SWR- meter at IC-718. **Table 5** shows SWR of the Beverage Antenna measured with help of MFJ-259B. **Table 5** shows SWR of the Beverage Antenna measured with help of internal SWR- meter at IC-718. Data for MFJ- 259B a little differ from data obtained with internal SWR- meter at IC-718. It happened because at antenna wire there is some stray RF voltage receiving from different radio stations. The RF voltage add some errors to the reading SWR by MFJ- 259B. In this case data obtained from the internal SWR- meter of IC-718 are close to the truth.

My Beverage Antenna has direction to Europe and East Cost of the USA. Antenna works perfect. Low bands 160 and 80- meter come to live. It was very good reception at all HF bands from 160- to 10 meter.

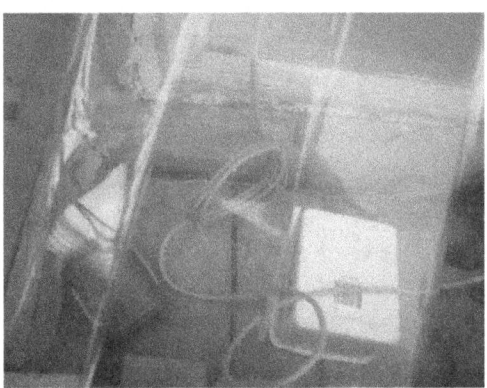

Figure 19 RF Choke near the Window

I could receive ham station from Europe and Asia from 160- to 10 meter. Of course, I cannot say that about transmitting operation on 160- 40- meter bands. It is noticeable that some power lost into the termination resistor. However, at good propagation the Beverage Antenna works perfect on transmission at all HF bands from 160- to 10 meter.

Table 5

Antenna Beverage with transformer made on LDG Balun Core. Load: 4.7-kOm x 11 Resistors (440 Ohm overall). Length of the 50- Ohm Coaxial Cable to antenna is 50 Feet. Measurement by MFJ- 259

Band	160	80	40	30	20	17	15	12	10
Z	75@j8	57@j4	42@j11	43@j19	50@j21	39@j28	24@j35	24@j12	97@j65
SWR	1.5	1.1	1.1	1.5	1.5	1.9	1.9	2.1	2.8

Table 6

Antenna Beverage with transformer made on LDG Balun Core. Load: 4.7-kOm x 11 Resistors (440 Ohm overall). Length of the 50- Ohm Coaxial Cable to antenna is 50 Feet. Measurement by SWR –meter of ICOM- 718

Band	160	80	40	30	20	17	15	12	10
SWR	1.7	1.0	1.0	1.5	1.1	1.8	1.8	2.0	2.9

Windom Compendium from RZ9CJ

By: Sergey Popov, RZ9CJ, Ekaterinburg, Russia
Credit Line: http://qrz-e.ru/forum/29-786-1

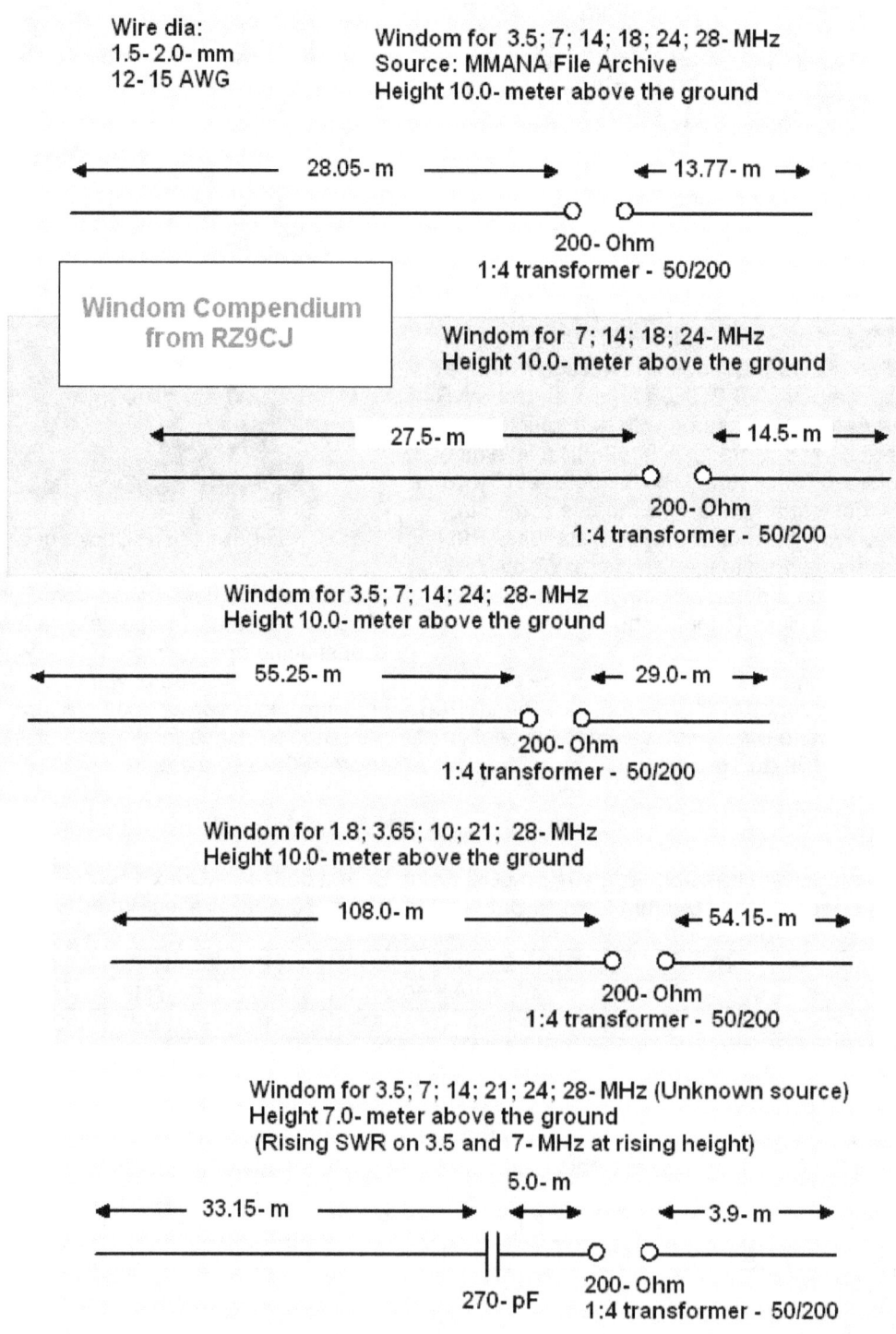

Windom UR0GT

The publication is devoted to the memory UR0GT.

Credit Line: Forum from:
www.cqham.ru

By: Nikolay Kudryavchenko, UR0GT

Windom is one of the oldest and reliable antennas that used in ham radio. There are lots modifications of Windom Antenna (or in other words Off Center Dipole Antenna). One of such modification was optimized by UR0GT. The antenna was optimized for 40, 20 and 10- meter bands. The main advantage of the Windom Antenna is that the antenna may work at several bands with low SWR.

The MMANA model of the Windom Antenna by UR0GT for 40, 20 and 10- meter Bands may be loaded: http://www.antentop.org/019/windom_ur0gt_019.htm

It is possible get by choosing the point of feeding of the antenna and lengths short and long parts of the dipole antenna. Height of the antenna as well influenced on the input impedance and resonance of the antenna. It would be useful to check the antenna in MMANA before antenna installation.

Simple broadband transformer 50/200- Ohm that may be used with the antenna described at Antentop- 01, 2015, pages 96- 97, *Broadband Transformer 50/200 Ohm by RZ9CJ.*

73! de UR0GT

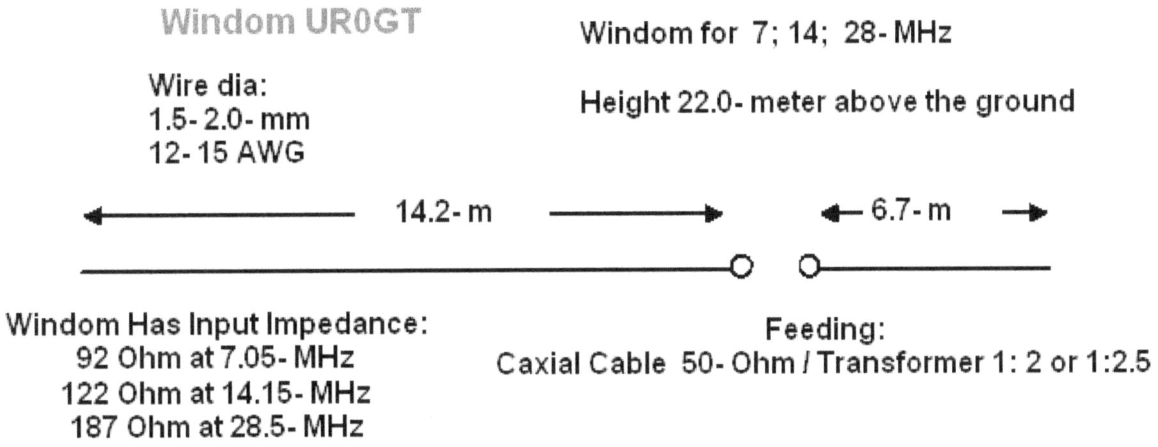

Windom UR0GT

Wire dia:
1.5- 2.0- mm
12- 15 AWG

Windom for 7; 14; 28- MHz

Height 22.0- meter above the ground

← 14.2- m → ← 6.7- m →

Windom Has Input Impedance:
92 Ohm at 7.05- MHz
122 Ohm at 14.15- MHz
187 Ohm at 28.5- MHz

Feeding:
Caxial Cable 50- Ohm / Transformer 1: 2 or 1:2.5

Figure 1 Windom Antenna by UR0GT for 40, 20 and 10- meter Bands

ANTENTOP

FREE e- magazine edited by hams for hams
Devoted to Antennas and Amateur Radio
www.antentop.org

www.antentop.org

Сервер Кубанских Радиолюбителей CQHAM.RU

www.cqham.ru

Two Vertical Antennas for 20-, 15- and 10- meter Bands

The publication is devoted to the memory UR0GT.

Credit Line: Forum from: www.cqham.ru

By: Nikolay Kudryavchenko, UR0GT

Below described two vertical antennas that work without any ATU at the 20-, 15 and 10- meter Bands. The antennas easy to made and easy to tune to the bands.

Figure 1 shows Triangle Vertical Antenna. **Figure 2** shows the antenna in 3d projection.

The MMANA model of the Triangle Vertical Antenna for 20, 15 and 10- meter Bands may be loaded: http://www.antentop.org/019/two_verticals_ur0gt_019.htm

Figure 3 shows Vertical Antenna with Inductor. **Figure 4** shows the antenna in 3d projection. The inductor has inductance in 10- microHenry. Mutual capacitance of the inductor should be not more the 1- pF. For example it is possible to use inductor with OD 4.8- cm, coiled by wire in 1-mm dia (19- AWG), gap between turns 1- mm, inductors has 15 turns.

The MMANA model of the Vertical Antenna with Inductor for 20, 15 and 10- meter Bands may be loaded: http://www.antentop.org/019/two_verticals_ur0gt_019.htm

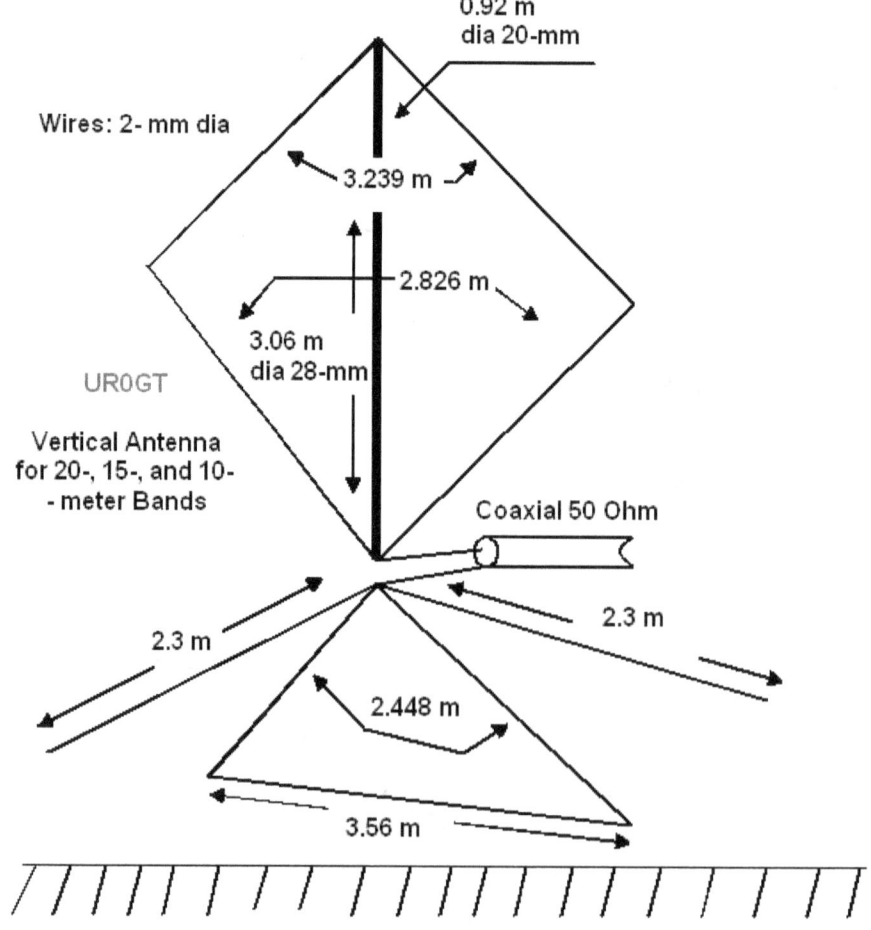

Figure 1 Triangle Vertical Antenna for 40, 20 and 10- meter Bands

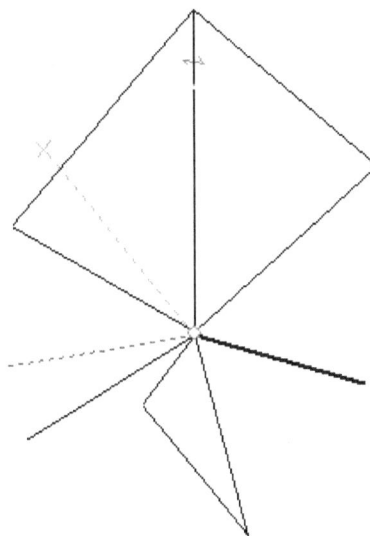

Figure 2 Triangle Vertical Antenna for 40, 20 and 10- meter Bands antenna in 3d projection

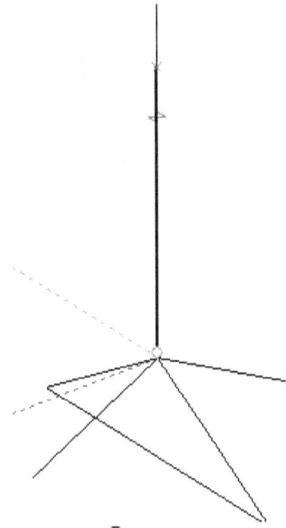

Figure 4 Vertical Antenna with Inductor for 40, 20 and 10- meter Bands antenna in 3d projection

Figure 5 shows impedance of the Triangle Vertical Antenna at 20- meter Band. Figure 6 shows SWR of the Triangle Vertical Antenna at 20- meter Band. Figure 7 shows DD of the Triangle Vertical Antenna at 20- meter Band. Figure 8 shows impedance of the Triangle Vertical Antenna at 15- meter Band. Figure 9 shows SWR of the Triangle Vertical Antenna at 15- meter Band. Figure 10 shows DD of the Triangle Vertical Antenna at 15- meter Band. Figure 11 shows impedance of the Triangle Vertical Antenna at 10- meter Band. Figure 12 shows SWR of the Triangle Vertical Antenna at 10- meter Band. Figure 13 shows DD of the Triangle Vertical Antenna at 10- meter Band. Antenna model was simulated at height 3- meter above the ground.

Figure 14 shows impedance of the Vertical Antenna with Inductor at 20- meter Band. Figure 15 shows SWR of the Vertical Antenna with Inductor at 20- meter Band. Figure 16 shows DD of the Vertical Antenna with Inductor at 20- meter Band. Figure 17 shows impedance of the Vertical Antenna with Inductor at 15- meter Band. Figure 18 shows SWR of the Vertical Antenna with Inductor at 15- meter Band. Figure 19 shows DD of the Vertical Antenna with Inductor at 15- meter Band. Figure 20 shows impedance of the Vertical Antenna with Inductor at 10- meter Band. Figure 21 shows SWR of the Vertical Antenna with Inductor at 10- meter Band. Figure 22 shows DD of the Vertical Antenna with Inductor at 10- meter Band. Antenna model was simulated at height 3- meter above the ground.

73! de UR0GT

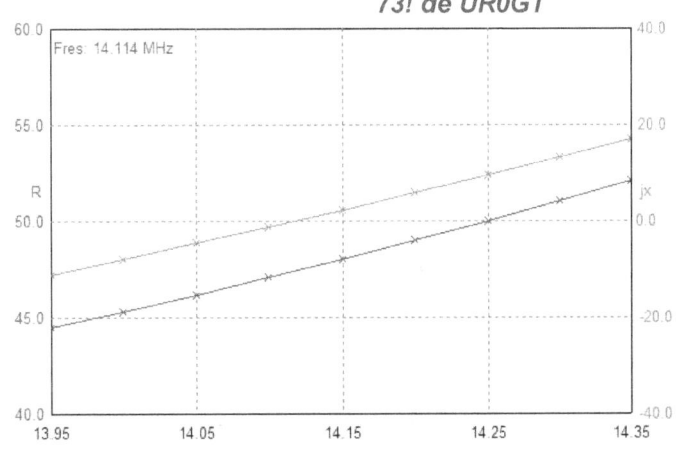

Figure 5 Impedance of the Triangle Vertical Antenna at 20- meter Band

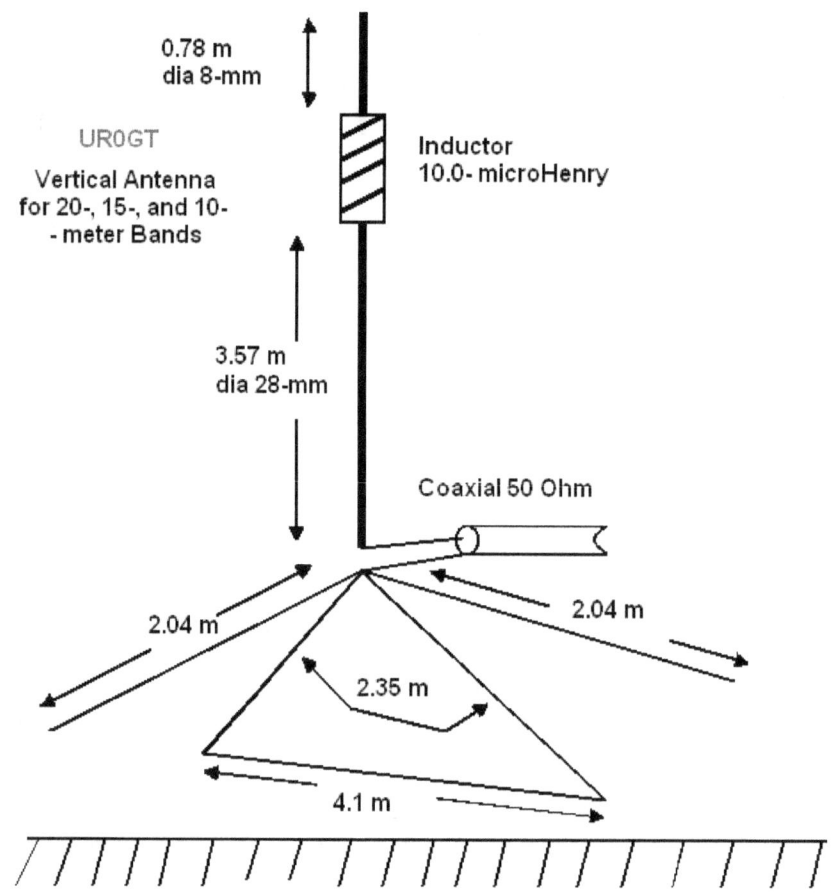

Figure 3 Vertical Antenna with Inductor for 40, 20 and 10- meter Bands antenna

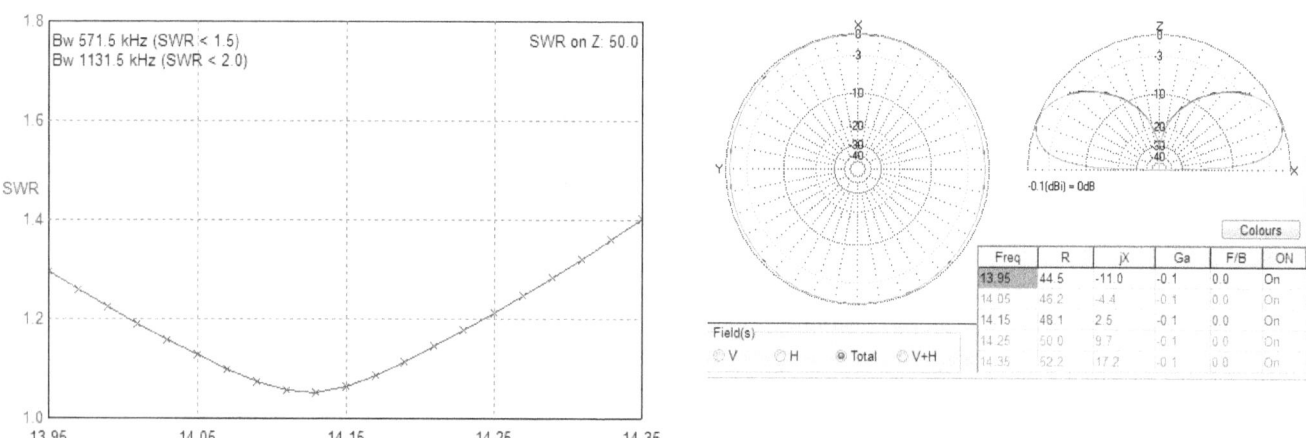

Figure 6 SWR of the Triangle Vertical Antenna at 20-meter Band

Figure 7 DD of the Triangle Vertical Antenna at 20-meter Band

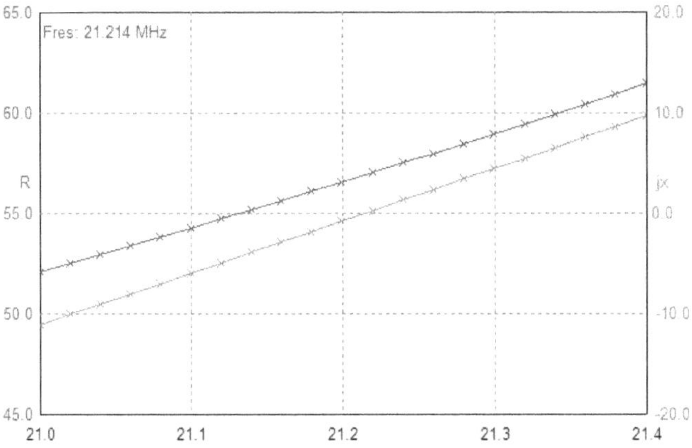

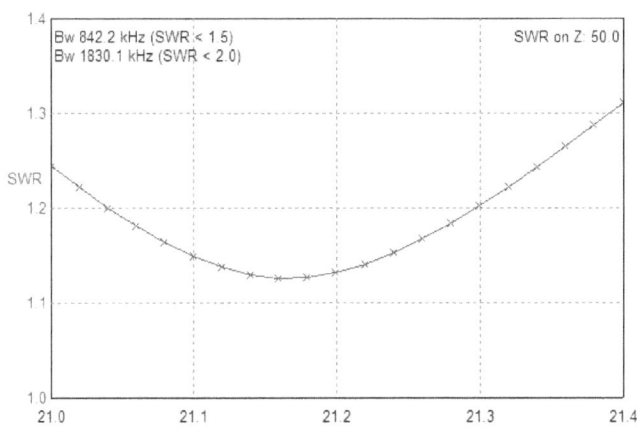

Figure 8 Impedance of the Triangle Vertical Antenna at 15- meter Band

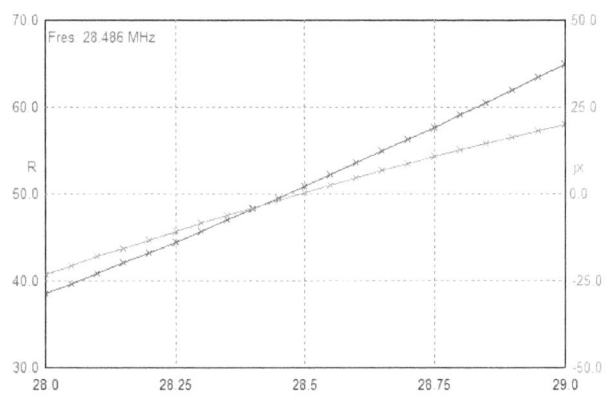

Figure 11 Impedance of the Triangle Vertical Antenna at 10- meter Band

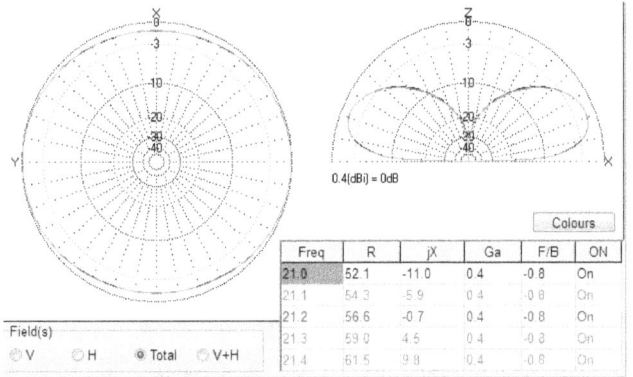

Figure 9 SWR of the Triangle Vertical Antenna at 15- meter Band

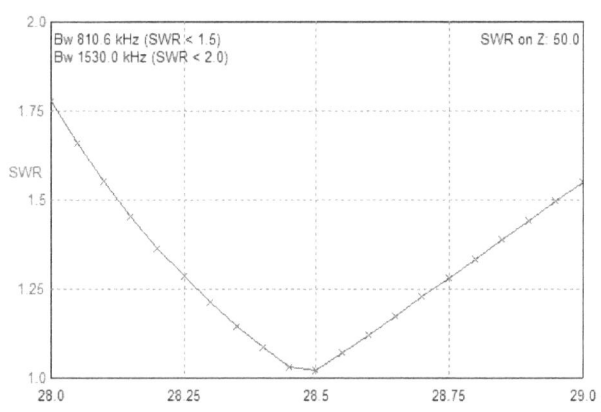

Figure 12 SWR of the Triangle Vertical Antenna at 10- meter Band

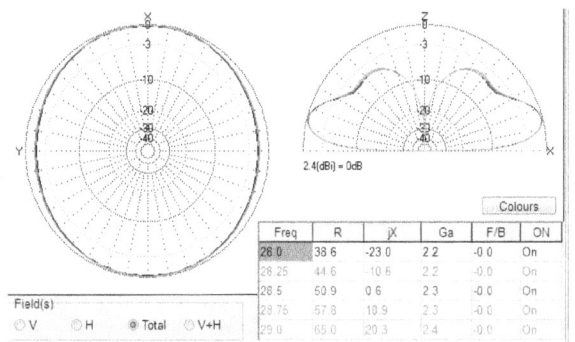

Figure 10 DD of the Triangle Vertical Antenna at 15- meter Band

Figure 13 DD of the Triangle Vertical Antenna at 10- meter Band

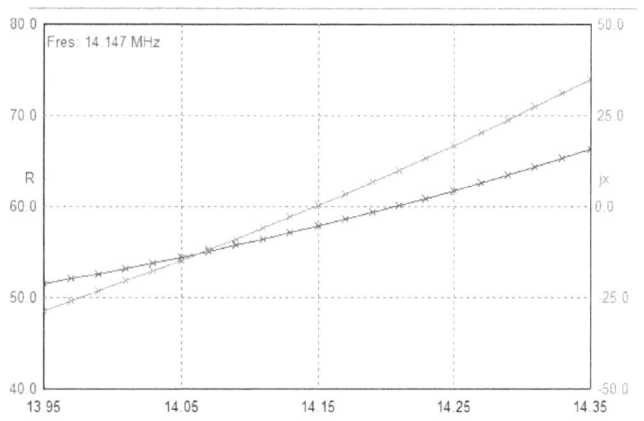

Figure 14 Impedance of the Vertical Antenna with Inductor at 20- meter Band

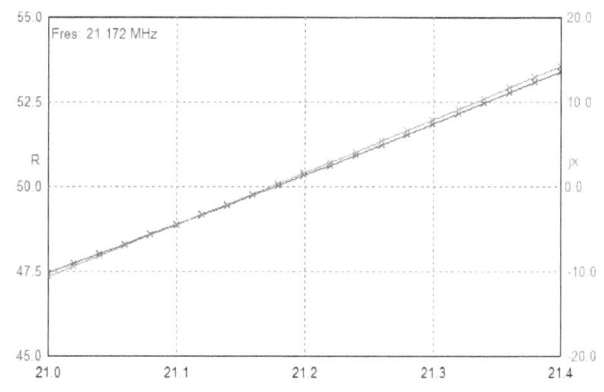

Figure 17 Impedance of the Vertical Antenna with Inductor at 15- meter Band

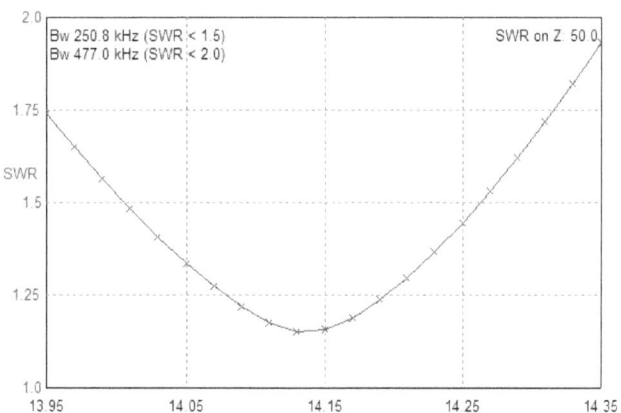

Figure 15 SWR of the Vertical Antenna with Inductor at 20- meter Band

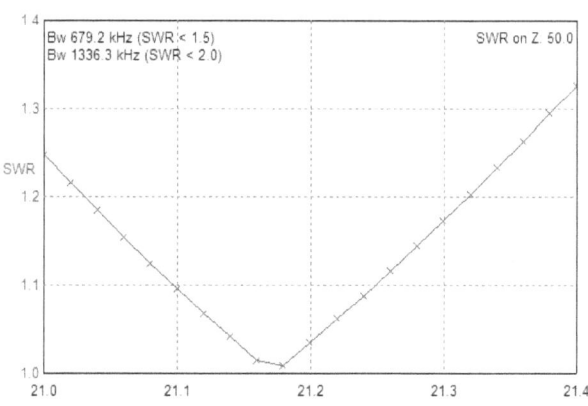

Figure 18 SWR of the Vertical Antenna with Inductor at 15- meter Band

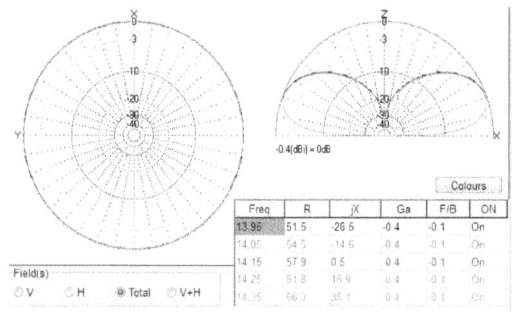

Figure 16 DD of the Vertical Antenna with Inductor at 20- meter Band

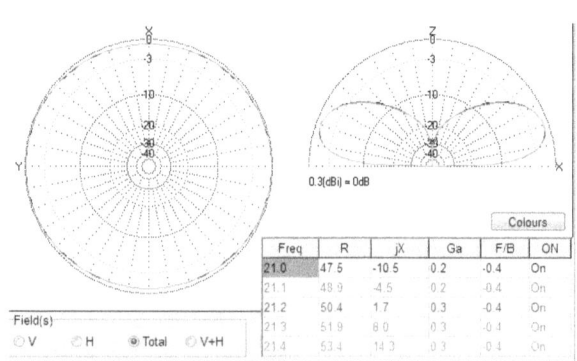

Figure 19 DD of the Vertical Antenna with Inductor at 15- meter Band

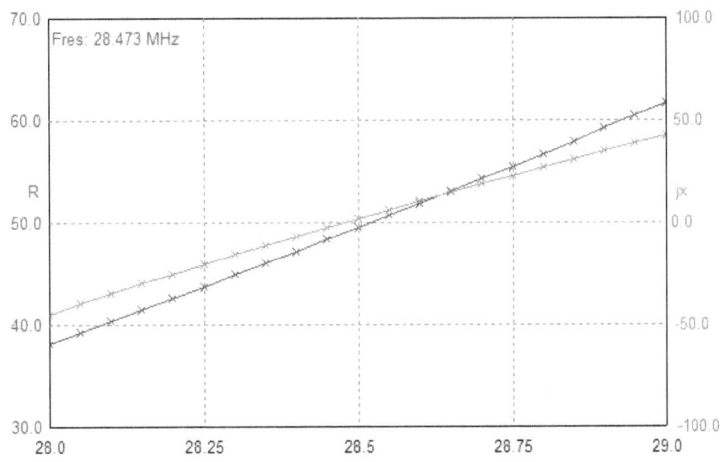

Figure 20 Impedance of the Vertical Antenna with Inductor at 10- meter Band

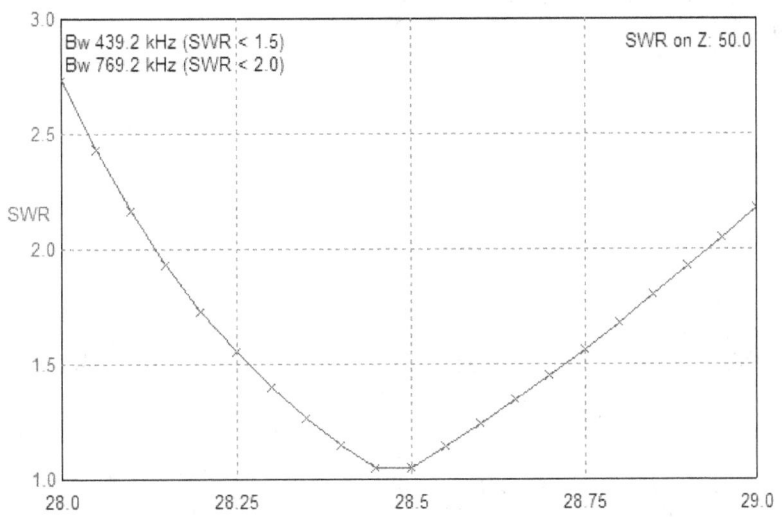

Figure 21 SWR of the Vertical Antenna with Inductor at 10- meter Band

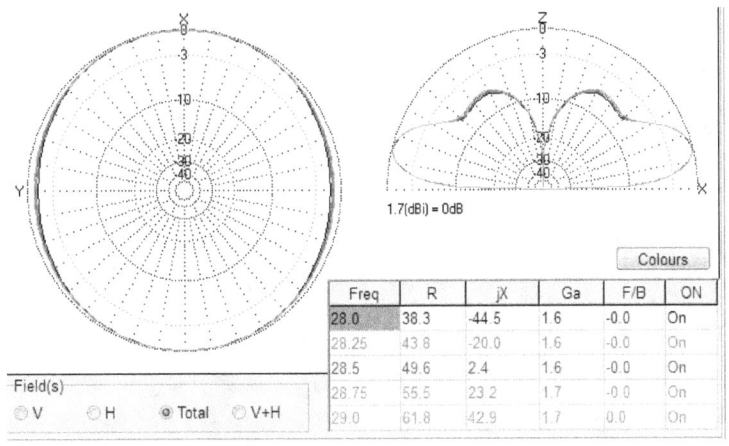

Figure 22 DD of the Vertical Antenna with Inductor at 10- meter Band

R3PIN Experimenters with UA6AGW Antenna

Aleksandr Grachev, UA6AGW

Credit Line: CQ-QRP # 48 (Autumn 2014), pp.: 19-22.

There are below described experimenters with UA6AGW Antenna made by Sergey Tetuyhin, R3PIN. Sergey would like create an UA6AGW Antenna for 2- meter Band. He did not have schematic of the antenna for 2- meter Band. He made two antennas that he believed would work at the 2- meter Band. However his attempt was not successful. But Sergey during the experimenters found some unusual sides at UA6AGW Antenna. Both antennas were made accordingly schematic shown at **Figure 1**.

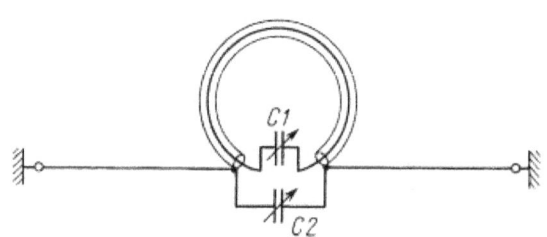

Figure 1 UA6AGW Antenna

Experiment # 1

Design of the Antenna

First antenna made by R3PIN is shown at **Figure 2**. The antenna was made as a table design. Height of the antenna was 45- cm. Loop of the antenna had diameter 13- cm. It was a copper tube in 8- mm OD. To make a "copper tube coaxial cable" inside of the tube was inserted plastic with central copper wire that was got from a piece of an old 50- Ohm coaxial cable. Plastic with central copper wire was in 2.5- mm diameter. Coupling loop was made from the same copper tube. Coaxial cable was soldered straight away to the coupling loop. Coupling loop had no electrical connection to the loop of the UA6AGW Antenna. **Figure 3** shows coupling loop at UA6AGW Antenna.

Antenna whiskers had telescopic design. The whiskers may be moved along the antenna loop by screw clamps. Capacitors C1 and C2 were Air – dielectric with capacitance 8- 140- pF. **Figure 4** shows the whiskers and capacitors.

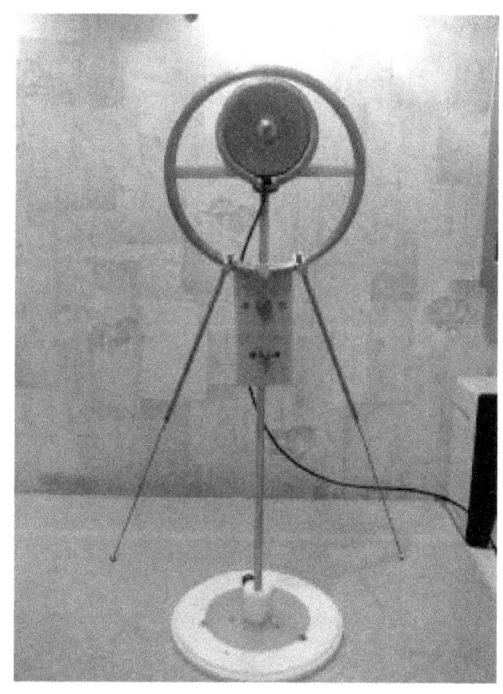

Figure 2 First Experimental Antenna made by R3PIN

Figure 3 Coupling Loop at UA6AGW Antenna

Data for the Antenna

Dimension of the antenna were too big for the 2- meter band. Antenna with help of C1 and C2 may be tuned across 18... 56- MHz. Test of the antenna was made at 10- meter band at frequency 28.850- MHz. Antenna was tested at position showed at **Figure 1**. Antenna had SWR 1.0: 1.0. C2 had maximum capacitance. Antenna was tuned to the 28.850- MHz by C1.

Antenna had Diagram Directivity similar to classical UA6AGW Antenna, i.e. the DD was ellipse sitting along whiskers of the antenna. Side suppression of the antenna was near minus15- dB. It was possible to make local QSOs when the antenna was placed at a table or windowsill. Experimenters with the antenna was loaded to Youtube at: http://www.youtube.com/watch?v=-uNrlRNcLu4&feature=em-upload_owner

Experiment # 2

Design of the Antenna

Second antenna made by R3PIN is shown at **Figure 5**. The antenna was made as a table design. Dimensions of the antenna were decreased compare to antenna from first experiment. Loop of the antenna has diameter 5.5- cm. It was a copper tube in 4.7- mm OD. To make a "copper tube coaxial cable" inside of the tube was inserted Teflon insulated wire in 0.27- mm diameter. Whiskers of the antenna had length 10- cm.

Figure 4 Whiskers and Capacitors

The whiskers made from strand tinned wire. Capacitors C1 and C2 were ceramic – dielectric with capacitance 6- 25- pF. Coupling loop was completely different from the counterpoint of antenna shown at **Figure 2**.

The coupling loop was placed athwart to radiation loop. Coupling loop had two turns of copper wire in 1.2- mm diameter (17-AWG). Gap between the turns was 5- mm. Coaxial cable was soldered directly to the coupling loop.

Two ferrite rings, one at the coupling loop another one at the connector side of the cable, were placed on to the coaxial cable. **Figure 6** shows the coupling loop. With help a fixture (similar to the clock hand) the loop could move along the radiation loop. **Figure 7** shows the fixture.

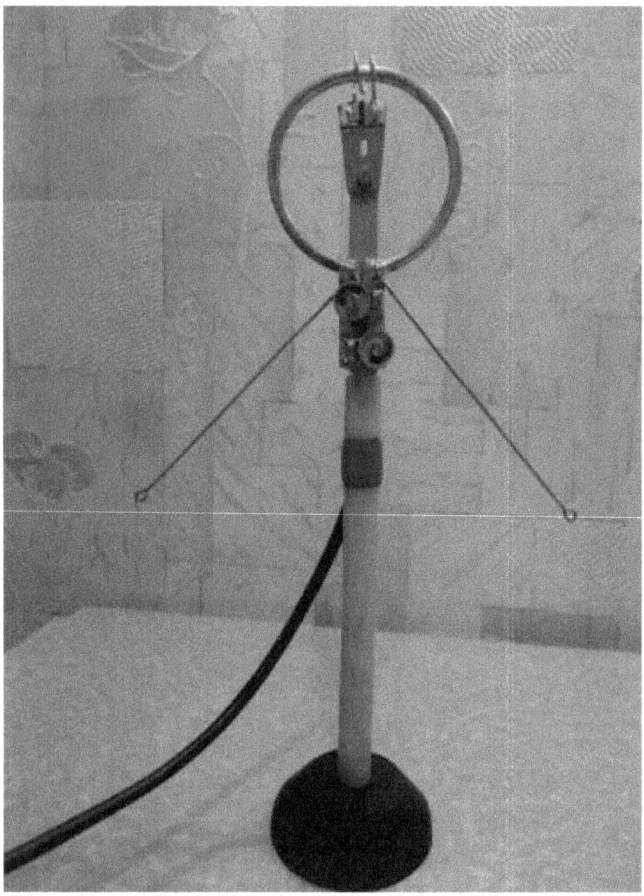

Figure 5 Small UA6AGW Antenna

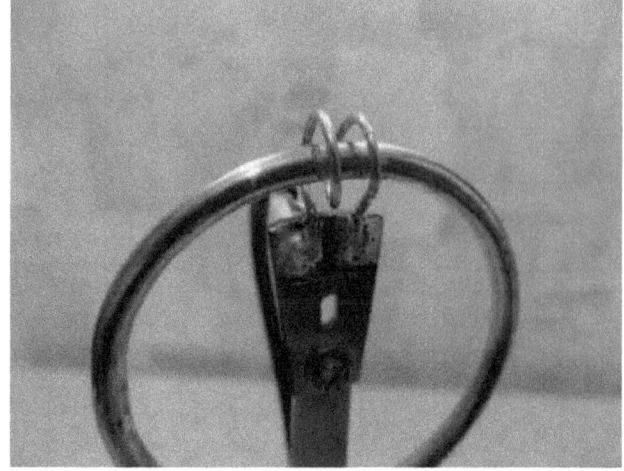

Figure 6 Coupling Loop

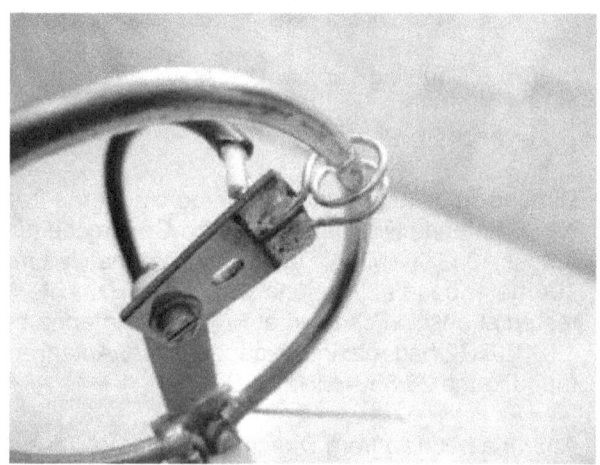

Figure 7 Fixture for Moving Coupling Loop

Data for the Antenna

Dimension of the antenna were too big for the 2- meter band. However the antenna with help of C1 and C2 may be tuned across FM Band 88… 108- MHz. Test of the antenna was made at the band only at receiving mode.

Antenna had one lobe diagram of directivity. The lobe was 30… 40 degree. Suppression from back and sides was at least minus 20- dB. Moving the coupling loop across the loop in one direction was going to increasing of the level receiving stations.

Moving the coupling loop across the loop in the opposition direction was going to decreasing of the level receiving stations and almost stopping of the receiving. Changing the coaxial cable connection to the coupling loop (visa versa) changed the direction. Looks like the one direction diagram directivity of the antenna was obtained due to the coupling loop placed athwart to radiation loop. Experimenters with the antenna was loaded to Youtube at: http://www.youtube.com/watch?v=qPBTSM-uFKI

Sergey, R3P9N, talking about his experimenters with UA6AGW Antenna

Dipole Antenna for 40- and 20- meter Bands

By: Vasily Perov, DL1BA (ex UK8BA)

Because I have no lots space at my backyard for antenna installation I like do experiments with shortened antennas. Below described one of my experimental shortened dipole antenna for 40- and 20- meter Bands. It takes for me only 1 and half hour for installation and tuning of the antenna. After that the dipole antenna was tested at CQ WW Contest (2015). I made 300 QSOs using 100- Wtts going into the antenna. **Figure 1** shows design of the antenna.

For the 20 meter Band the antenna is a full sized lambda/4 (each side) dipole antenna. For the 40- meter band the antenna is a shortened dipole antenna. I used simulation program NEC2 to create the antenna. However on practice the sizes were a little different to the theoretical ones.

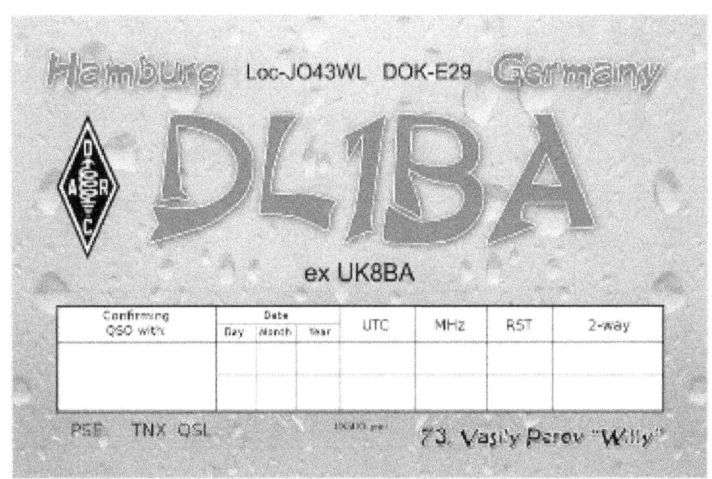

DL1BA QSL Card

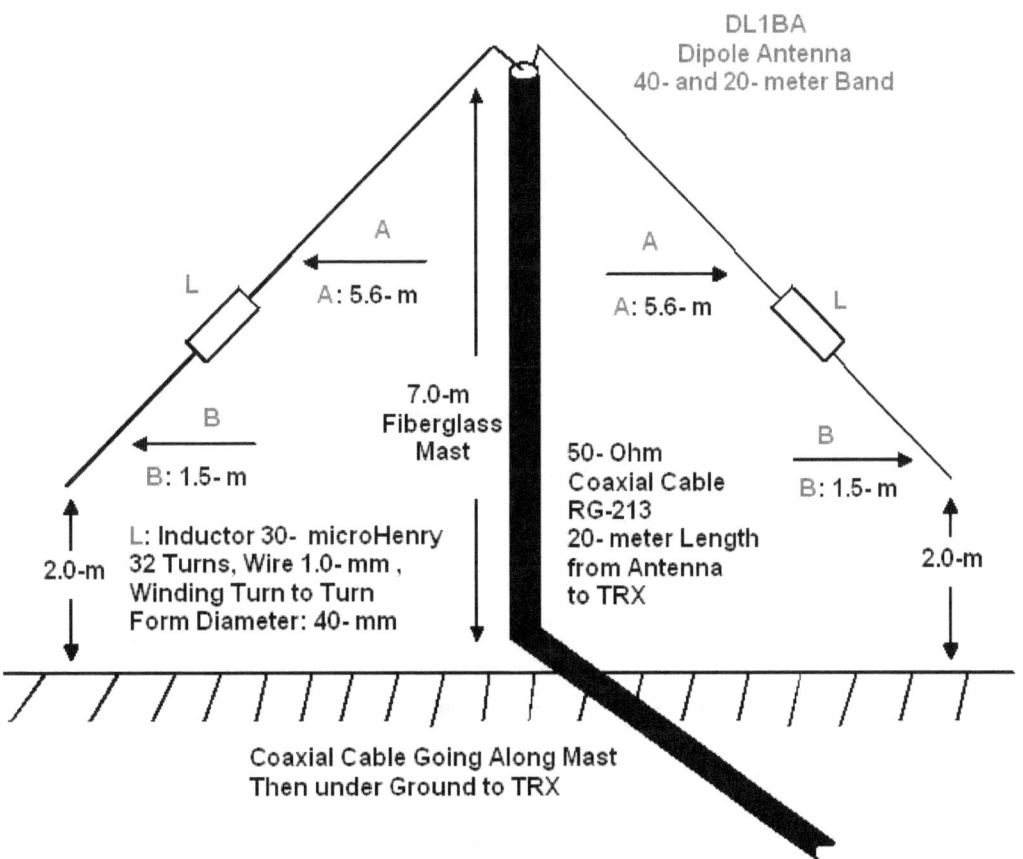

Figure 1 Dipole Antenna for 40- and 20- meter Bands

Antenna dimensions are depended on the placement and inductance of the inductor. However the inductance of the inductor is not critical for antenna operation. Most important thing is that the inductors have the same inductance. In my case the inductors (measured by me) have inductance 31.4- micro Henry. The inductance is influenced to the parameters of the antenna. The more it is inductance of the inductor the less length of the parts of the antenna and the less passband of the antenna.

Antenna was placed at the mast in 7- meters height. The mast was made from a fiberglass plastic. Antenna was made from stranded wire in plastic insulation. Antenna was fed through 50- Ohm coaxial cable. the antenna.

I have used RG 313 type. This cable was going along the mast and then going underground to my shack. Length of the cable was near 20- meter. Weather practically did not influence to the antenna.

Antenna was easy tuned to SWR 1:1 at both Bands. Antenna had bandpass near 200- kHz (at SWR 1.5:1) at the 20 meter Band and bandpass near 75- kHz (at SWR 1.5:1) at the 40 meter Band. At first step the antenna should be tuned at the 20- meter Band by symmetrical changing lengths of the parts A. Then antenna should be tuned at the 40- meter Band by symmetrical changing parts B. Changing the length of the parts B practically did not influenced to the operation at the 20- meter Band.

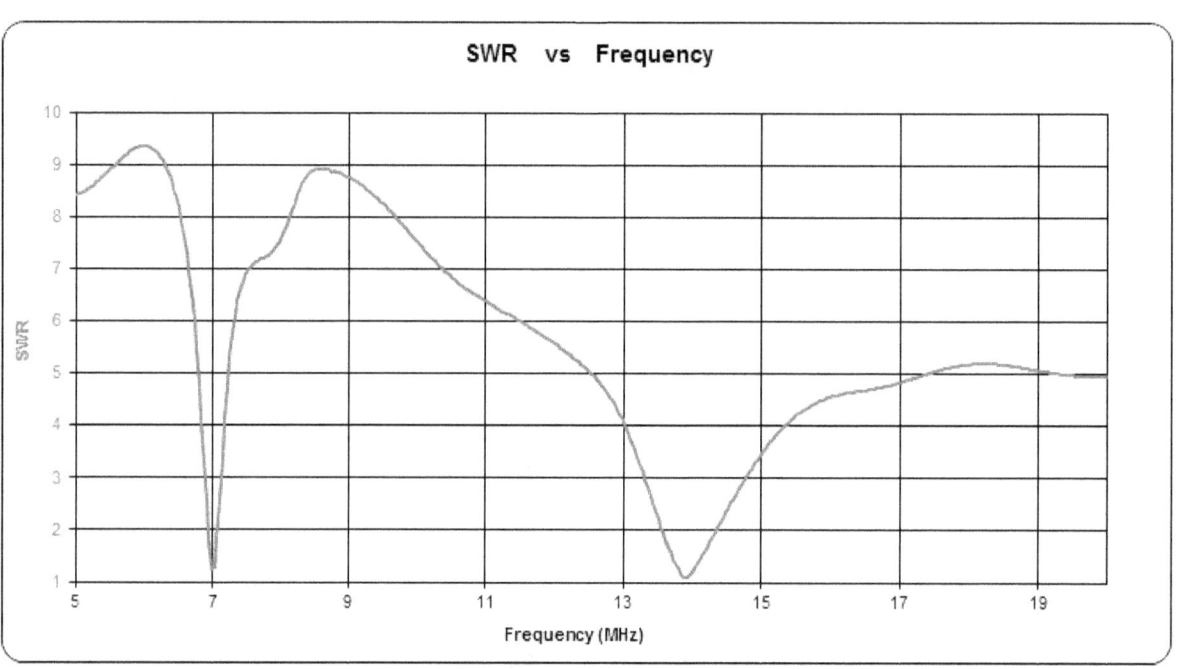

Figure 2 SWR of the Dipole Antenna for 40- and 20- meter Bands from 5 to 19.5- MHz

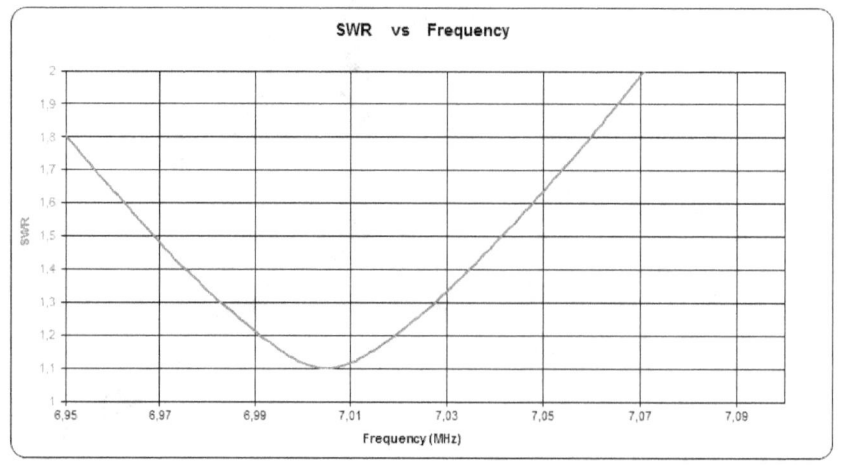

Figure 3 SWR of the Dipole Antenna for 40- and 20- meter Bands at the 40- meter Band

In my case changing length B on to 1- cm caused changing resonance frequency of the antenna up to 100- kHz at the 40- meter Band. **Figure 2** shows SWR of the antenna from 5 to 19.5- MHz. The plot was taken with VNA N2PK.

Figure 3 shows SWR of the antenna at the 40- meter Band. **Figure 4** shows SWR of the antenna at the 20- meter Band. SWR was taken at the end of the coaxial cable going from antenna. The program Zplots where the VNA is working has function to remove length of cable (going to antenna) from the antenna measurements.

Figure 5 shows SWR of the antenna from 5 to 19.5- MHz with turned on function "remove coaxial cable length." As you can see the plot at the bands of the interest is practically identical to plot shown at **Figure 2**.

The MMANA file of the Triangle Vertical Antenna for 20, 15 and 10- meter Bands may be loaded: http: // www.antentop.org/019/dl1ba_dipole_019.htm The file was prepared by RW4HFN. Simulation the Triangle Vertical Antenna for 20, 15 and 10- meter Bands in MMANA shows very close theoretical parameters of the antenna to practical obtained ones.

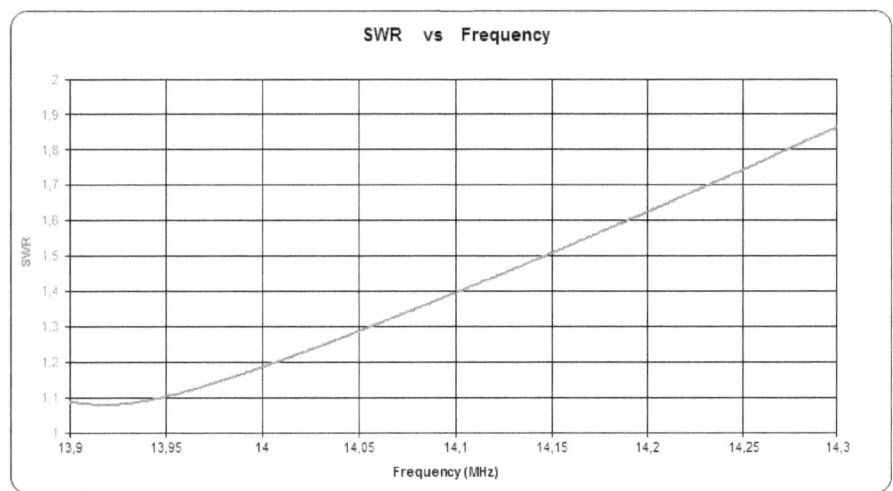

Figure 4 SWR of the Dipole Antenna for 40- and 20- meter Bands at the 20- meter Band

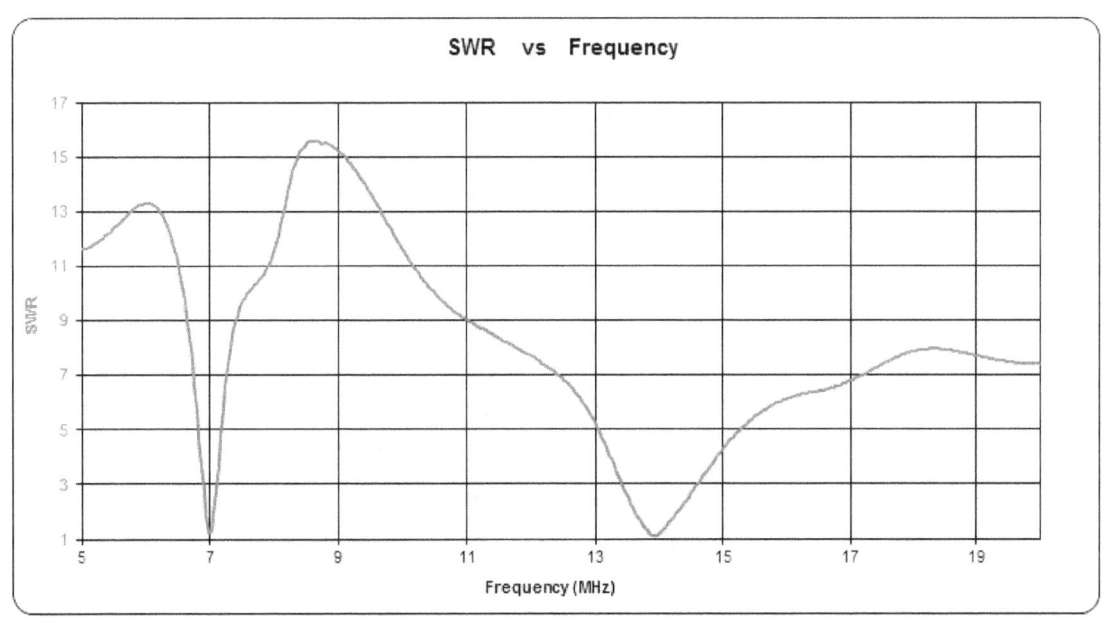

Figure 5 SWR of the Dipole Antenna for 40- and 20- meter Bands from 5 to 19.5- MHz with turned on function "remove coaxial cable length"

Modified Dipole Antenna DL1BA for 40- and 20- meter Bands

By: *Igor Vakhreev, RW4HFN*

In my opinion the explanation how the DL1BA Antenna (*Antentop 01- 2015, pp: 53-55, Dipole Antenna for 40- and 20- meter Bands*) is working at the 20- meter Band is very simple. Parts of the antenna- there are long wire (5.6-meter length) before inductor, inductor and short wire (1.5- meter length) after inductor – make 1.5- lambda dipole at the 20- meter Band. **Figure 1** shows current distribution in the DL1BA Antenna at the 20- meter Band. The current distribution proves my suggestion.

As you can see from the **Figure 1** the minimal current (current node) is placed at half meter up from the inductor. It allows find another approach for tuning of the DL1BA Antenna. It is possible make one side of the antenna a little short- say to 30... 50- cm.

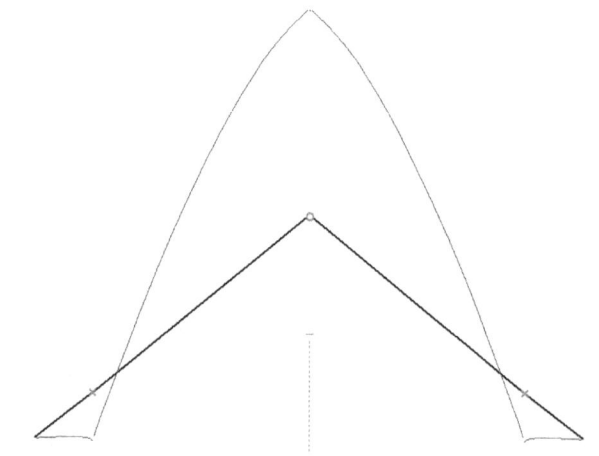

Figure 1 Current Distribution in the DL1BA Antenna at the 20- meter Band

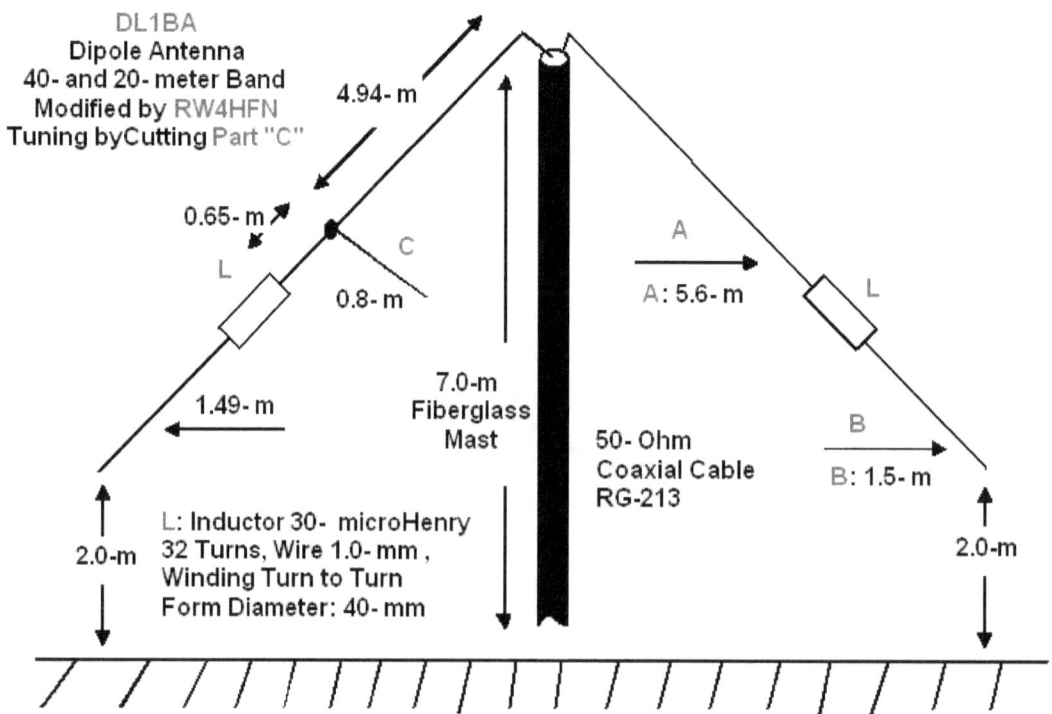

Figure 2 Modified Dipole Antenna DL1BA for 40- and 20- meter Bands

At the current node at the short side it is added wire in length (what the antenna is shortened) 30- 50- cm. At this case the adjusting of the antenna would be simple. At the 20- meter band the antenna is tuned by shortening the added wire. At the 40- meter Band the antenna is tuned by shortening of the one short wire after inductor that laid at non modification (that is without added wire) side. Both tuning practically do not influenced to each other. **Figure 2** shows the modified antenna.

The MMANA file of the Modified Dipole Antenna DL1BA for 40- and 20- meter Bands may be loaded: http://www.antentop.org/019/rw4hfn_dipole_019.htm Simulation the Antenna in MMANA shows that it is possible to tune the antenna for both Bands by playing the lengths of the added wire (part C see **Figure 2**) and the short wire after inductor (part B at right side antenna shown on **Figure 2**).

Figure 3 shows SWR of the modified antenna at the 40- meter Band. **Figure 4** shows SWR of the modified antenna at the 20- meter Band.

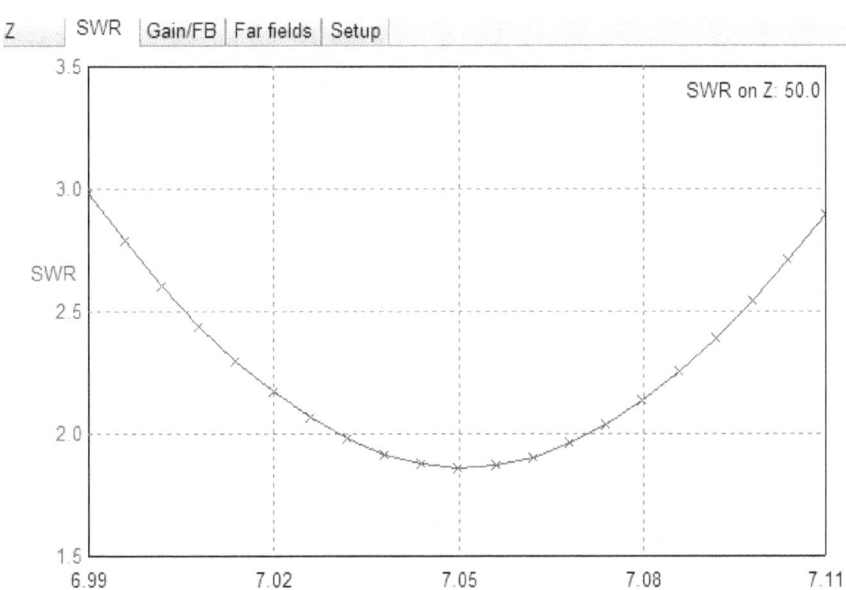

Figure 3 SWR of the Modified DL1BA Antenna at the 40- meter Band

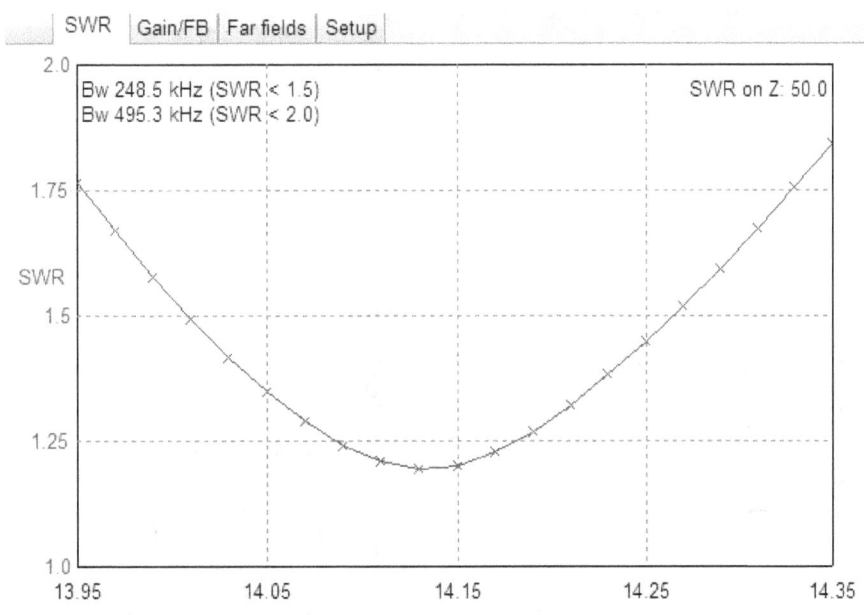

Figure 4 SWR of the Modified DL1BA Antenna at the 20- meter Band

Modified DL1BA Dipole Antenna for 40-, 20- meter Bands with additional 10- or 15- meter Band

By: *Igor Vakhreev, RW4HFN*

DL1BA Antenna (*Antentop 01- 2015, pp: 53-55, Dipole Antenna for 40- and 20- meter Bands*) may be modified for working at additional 10- or 15-meter Band. **Figure 1** shows design of the antenna.

As you can see from the **Figure 1** parts for additional band are added into the antenna design.

The MMANA file of the Modified DL1BA Dipole Antenna for 40- and 20- meter Bands with additional 10- or 15- meter Band may be loaded: http: // www.antentop.org/019/rw4hfn_modified_dipole_019.htm

Simulation of the antenna in MMANA showed that the antenna may be tuned at resonance at all of three Bands.

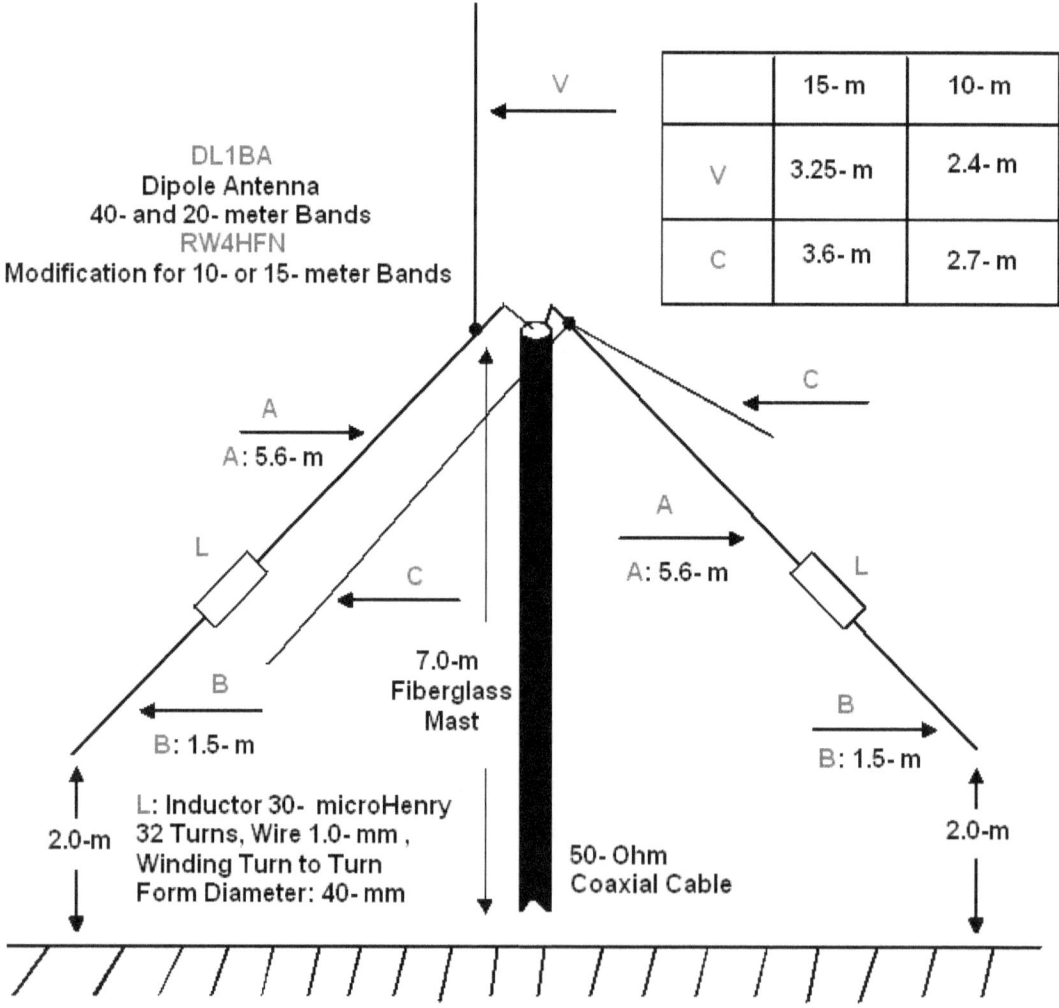

Figure 1 Modified DL1BA Dipole Antenna for 40- and 20- meter Bands with Additional 10- or 15- meter Band

Modified DL1BA Dipole Antenna for 40-, 20-, 15-, and 10- meter Bands

By: *Igor Vakhreev, RW4HFN*

DL1BA Antenna (*Antentop 01- 2015, pp: 53-55, Dipole Antenna for 40- and 20- meter Bands*) may be modified for working at additional 10- and 15- meter Bands. **Figure 1** shows design of the antenna.

As you can see from the **Figure 1** parts for additional bands are added into the antenna design.

The MMANA file of the Modified DL1BA Dipole Antenna for 40-, 20-, 15-, and 10- meter Bands may be loaded: http://www.antentop.org/019/rw4hfn_modified_dl1ba_019.htm

Simulation of the antenna in MMANA showed that the antenna may be tuned at resonance at all of the four Bands.

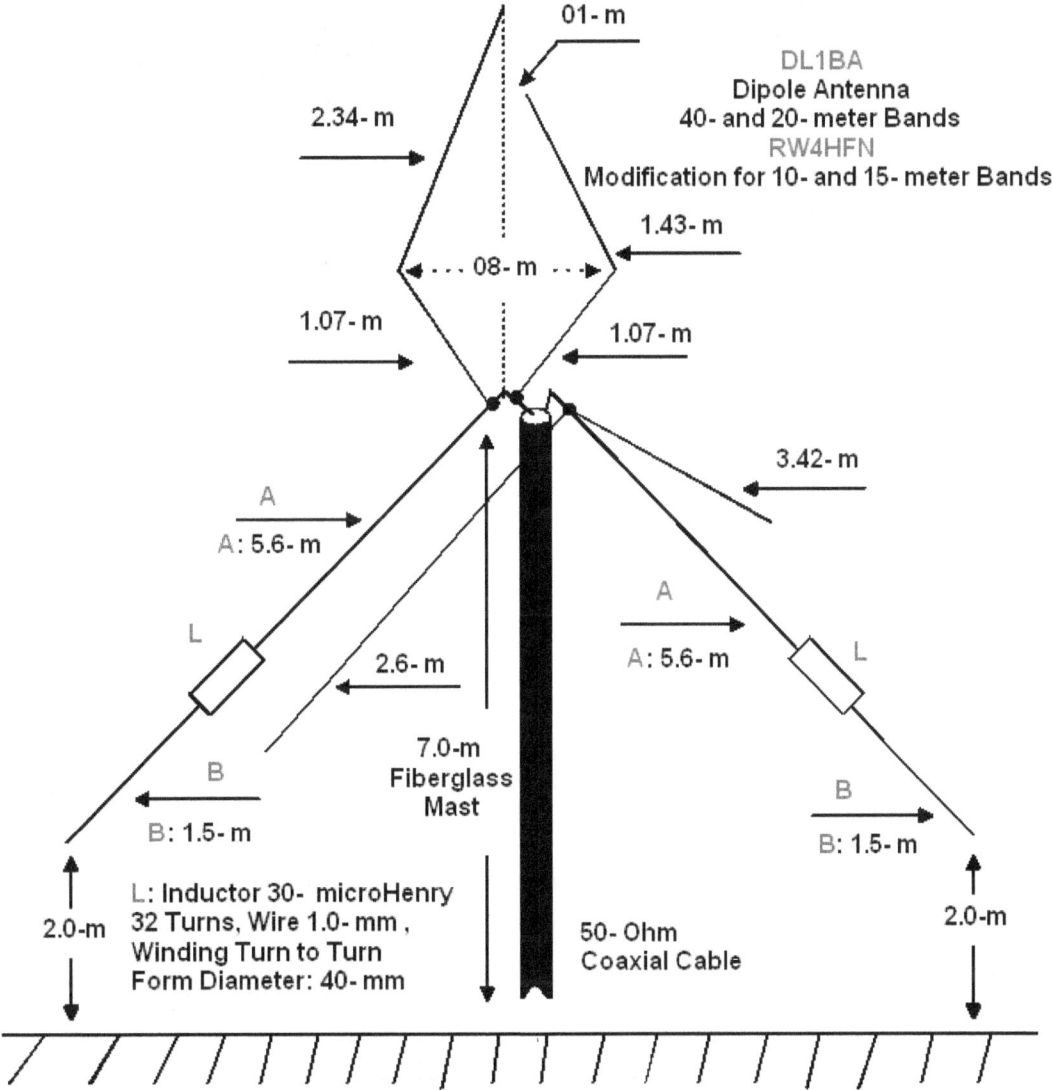

Figure 1 Modified DL1BA Dipole Antenna for 40-, 20-, 15-, and 10- meter Bands

Antenna for 80-, 40-, and 15- meter Band

By: *Vladas Zhalnerauskas, UP2NV*

W3DZZ Antenna is widely used among radio amateurs. The antenna shows the efficiency at several bands (as usual from 3 to 5) at minimal stuff to do it. The described below antenna is a modification of the W3DZZ Antenna. The antenna could work at 80- 40- and 15- meter Bands. **Figure 1** shows all steps to transform W3DZZ Antenna to my design.

It is known that classical W3DZZ Antenna has traps for one or several bands. In my case the traps were changed at first to quarter wave stubs then the stubs were superposed with antenna wires. I got compact and reliable antenna that worked good at three bands- 80-, 40-, and 15- meters. Sizes of the antenna at **Figure 1** are given in centimeters. Stubs were fixed with help of dielectric struts. Antenna should feed through 75- Ohm coaxial cable if it is placed at some height above the ground. If the antenna made in the shape of inverted V the 50- Ohm coaxial cable should be used for feeding the antenna. Antenna was made from 2- mm (12- AWG) copper wire. Coaxial cable should be placed at right angle to the antenna wires.

The Antenna was used for several years at UP2NV station and it was showed a good result.

73! de UP2NV

UP2NV. Klaipeda, 1981.
Credit Line:
http://deltaclub.org.ua/bibliteka/history-taught-extramural-competitions-hf-year-1981.html

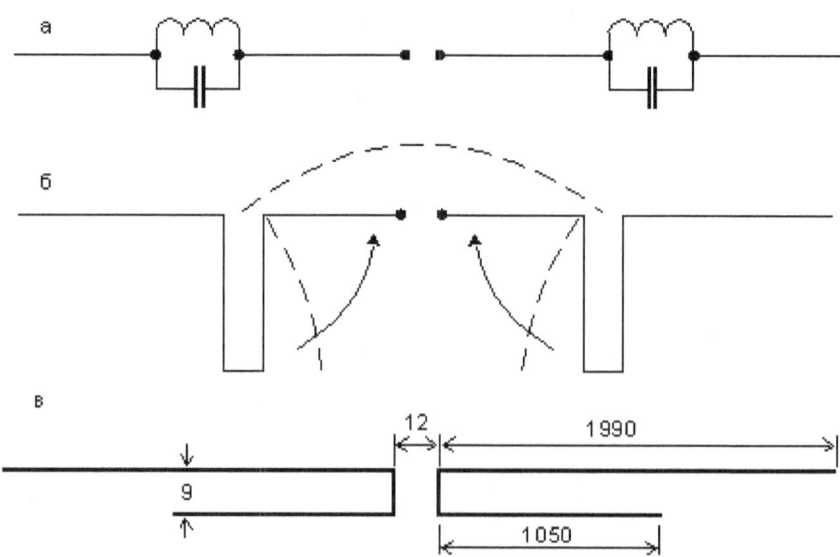

Figure 1 Transformation W3DZZ Antenna to UP2NV Antenna

ANTENTOP — Antennas for 50- and 70- MHz

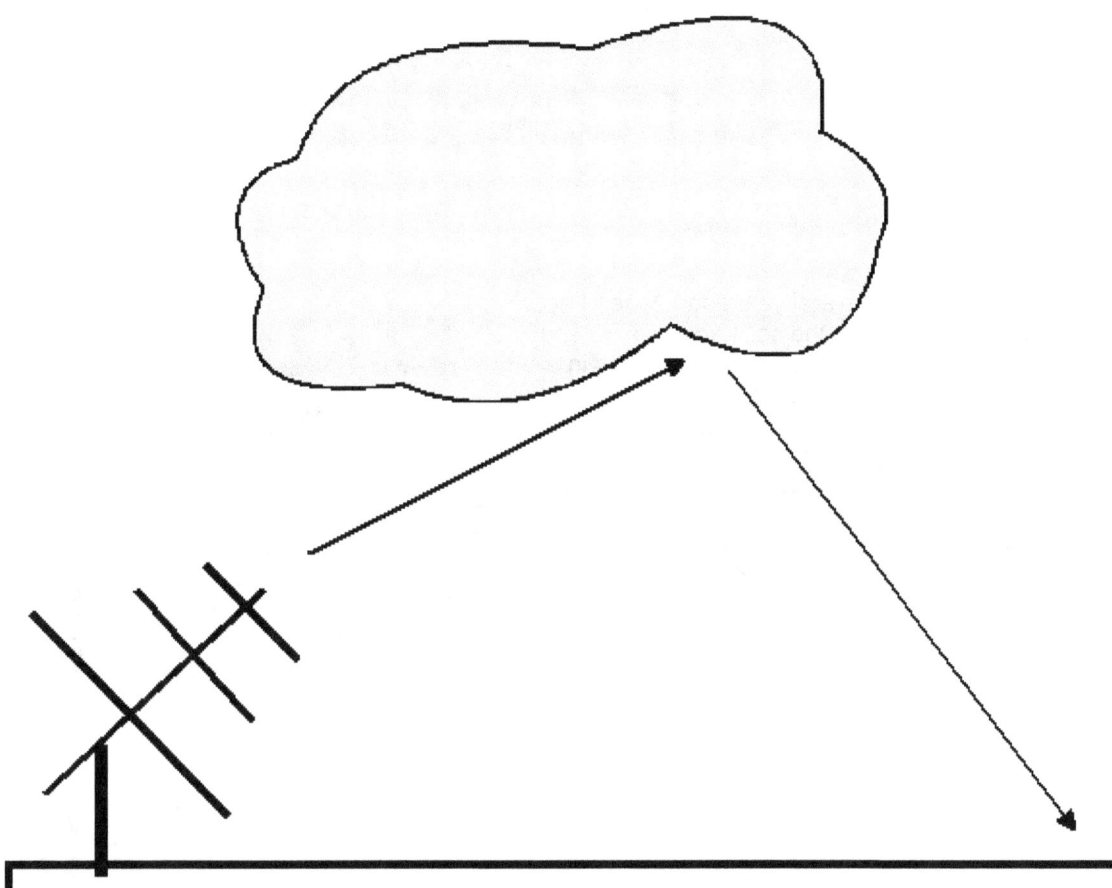

Antennas for 50- and 70- MHz

www.antentop.org

Antenna for 50 and 70- MHz Band

By: Alex Karakaptan, es4uy, uy5on

The antenna was designed several years ago, when I got Estonian call sign ES4UY. With the call I able use 50 and 70- MHz bands from Estonia. Need to say that I was in a very rare square- KO49CJ. So I need an antenna for those bands. Restricted place could not allow me to create something serious. So, wire in length in 3.75 meter was placed on to perimeter of my window then the wire was fixed at balcony. The wire was connected to the home brew ATU. Counterpoise in length 0.7- meter was connected to the ATU. **Figure 1** shows the antenna on the window. **Figure 2** shows design of the antenna. **Figure 3** shows schematic of the ATU. **Figure 4** shows design of the ATU.

ES4UY QSL Card

Figure 1 Antenna for 50 and 70- MHz Band Placed on the Window

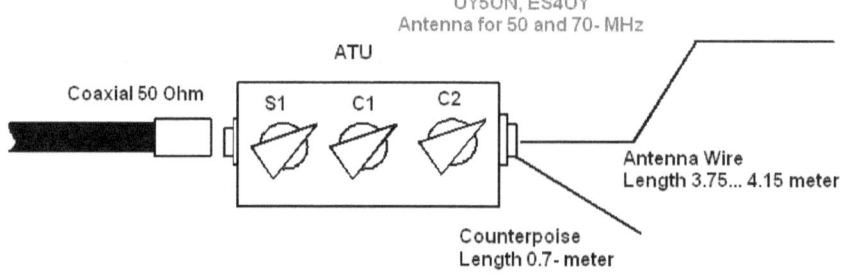

Figure 2 Design of the Antenna for 50 and 70- MHz Band

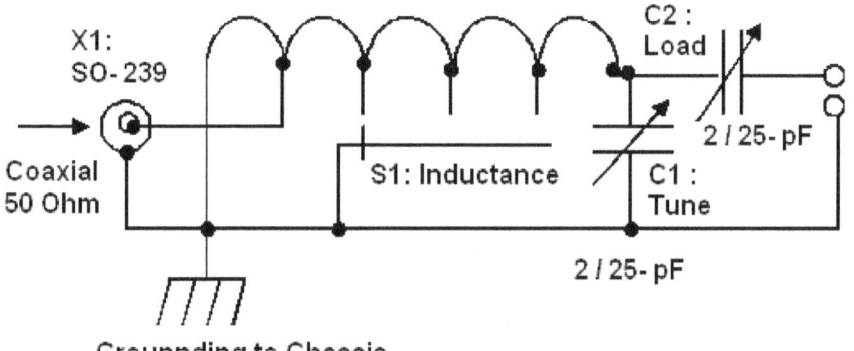

Figure 3 Schematic of the ATU

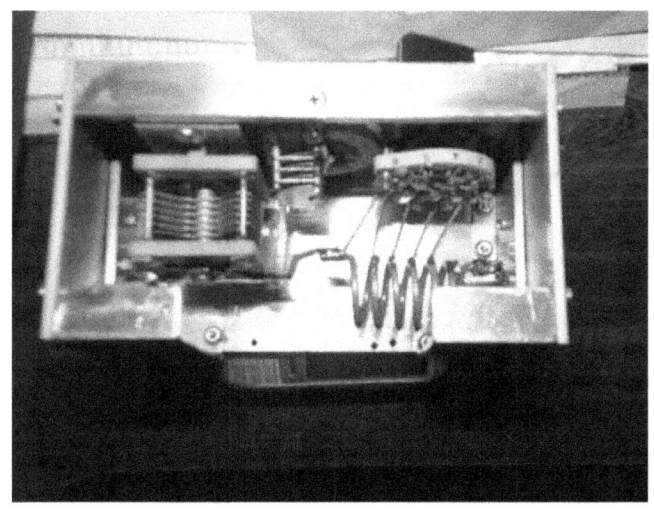

Figure 4 Design of the ATU

The antenna is kind of Windom Antenna for 50- and 70- MHz. Length in 3.5 is a long part of the antenna counterpoise in 0.7- meter is short part of the antenna. ATU match the antenna with coaxial cable. I made ATU inside of aluminum box. Inductor was wound on a form with OD- 25- mm and contained 5 turns. Inductor was coiled by silver- plated wire in diameter 1.8- mm (13- AWG).Taps were made from 1, 2, 3, and 4-th turn.

First tap of the inductor through a short length connected to the "hot" lead of the RF socket X1. Beginning of the inductor soldered to the ground of the RF socket X1. Taps 2, 3, 4 connected by a short length to the switch S1.

Capacitors C1 and C2 were hi- Q air capacitors. Capacitor C2 insulated from the box. I managed tune capacitors of the ATU in such way that I could change working Band only with switch S1. ATU has losses only minus 0.53 dB. It was checked by ADVANTEST made analyzer.

Below you can see data from my Log for 25- 26-June, 2015. There are 29 QSOs and one QSO with CT1HZE- distance 3600- km. So the antenna works and works well. It should be used at any place with restricted conditions.

QSO by ES4UY on 70mhz
Type of propagation: ES
Loc: KO49CJ
TRX: Ft-847 -60 wats Ant: Home Made with 3.75m Window Antenna.
Format: Call, Loc, RS/RST

DJ7MN, JN58WH, 599/ 599:   S57LM, JN76HD, 599/559;   OK2BGW, JN89CH, 599/ 599; OZ3ZW, JO54RS, 59/ 59;   OK2BGW, JN89CH, 57/ 57;   OK2BRD, JN99ET, 59/ 59; PA0RDY,JO22KJ, 579/ 559; PA0RDY, JO22KJ,579/ 559; OM5KM, JN98BG, 559/599; HA3GR, JN86VK,579/ 599; PA2IP, JO23VF, 599/599; SP3RNZ, JO92DF, 559/ 559; LZ1AG,KN22ID, 559/ 559; OK1KT, JO80CH, 599/599; HA3GR, N86VK, 599/ 599;   SP6GWB, JO80JG, 599/ 599; YO7BSN, KN15OA, 599/ 599; 9A2SB, JN95GM, 559/559; CT1HZE, IM57NH,55/55; PG5V, JO21, 559/ 549 ; PA0O, JO33HG, 599/599; PA2M, JO21IP,59/59; ON4PS, JO20KQ, 599/599; ON4PS, JO20KQ,599/ 599; DK2PH, JO41GV, 599/559; OZ1BNN, JO55PM, 55/ 55; OZ2OE, JO45VV, 59/ 59;   OZ3ZW, JO54RS, 59/ 59;   OZ8ZS, JO55RT, 57/59; OZ1JXY, JO46TX,57/ 55.

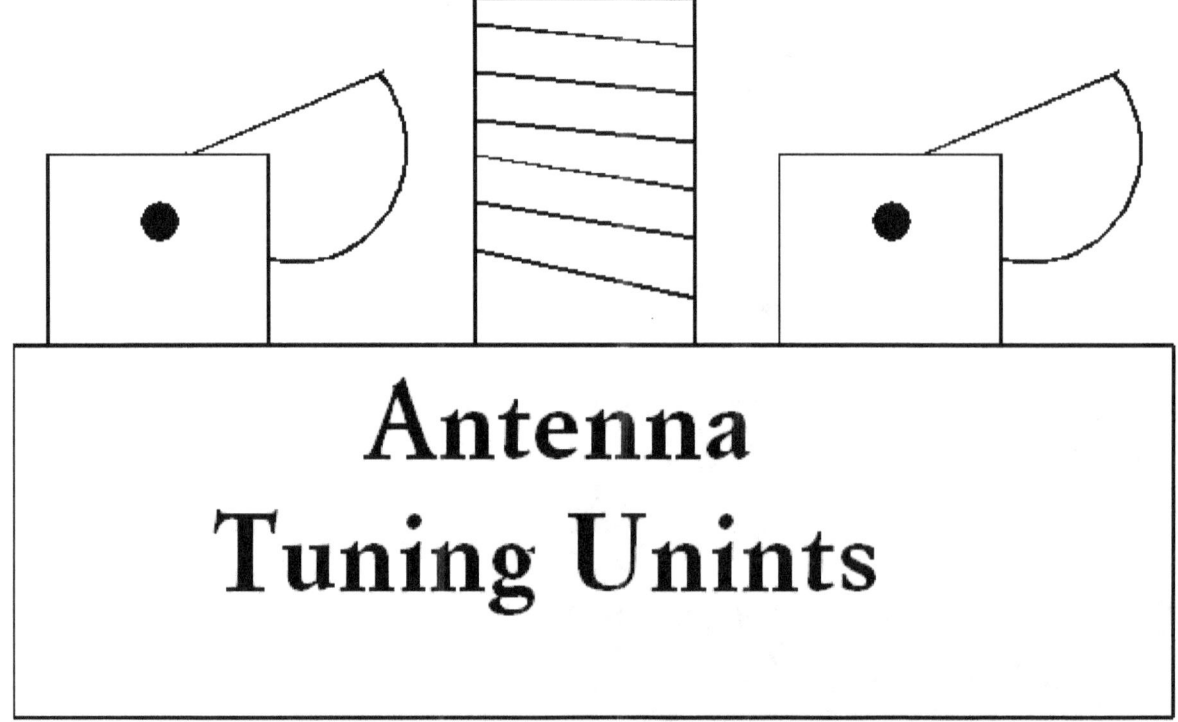

Symmetrical ATU

By: Vasily Perov, DL1BA (ex UK8BA)

Prototype of the tuner was made by VK5RG. The tuner was found by me at „Das DARC Antennenbuch"(*Reference 1*). However at the book there was given only brief description of the unit. The tuner takes my attention and by trial-and-error method I found the design (data for Inductors and Capacitors) of the tuner. **Figure 1** shows schematic of the Symmetrical ATU. Pay attention that at the tuner the rotary switches S1-1 and S1- 2 do shortening of the unused turns.

The Symmetrical ATU is kind of usual two Pi- circuits, C1 is capacitor at hot end- capacitor that tune the inductor L2 to resonance and C2 is capacitor at cold end- capacitor that does matching for the load. **Figure 2** shows design of the Symmetrical ATU.

All three inductors are placed in row. The inductors are wound by copper wire in 1.5- mm diameter (15- AWG). Inductor L1 contains 4 turns and placed between matching inductors L2-1 and L2-2. Inductors wound on a dielectric plate (PCB plate without foil) by dimensions – 150x 80x 2mm. Two row holes were drilled in the plate. The rows were 50 – mm apart and it was 3- mm distance between the holes.

At first step the inductors were being wound on to a form in 50- mm diameter. Then the dummy inductor was taken off from the form and inserted into the plate turn by turn. Inductor L1 contains 4 turns. Inductors L2- 1 and L2- 2 have 16 turns each.

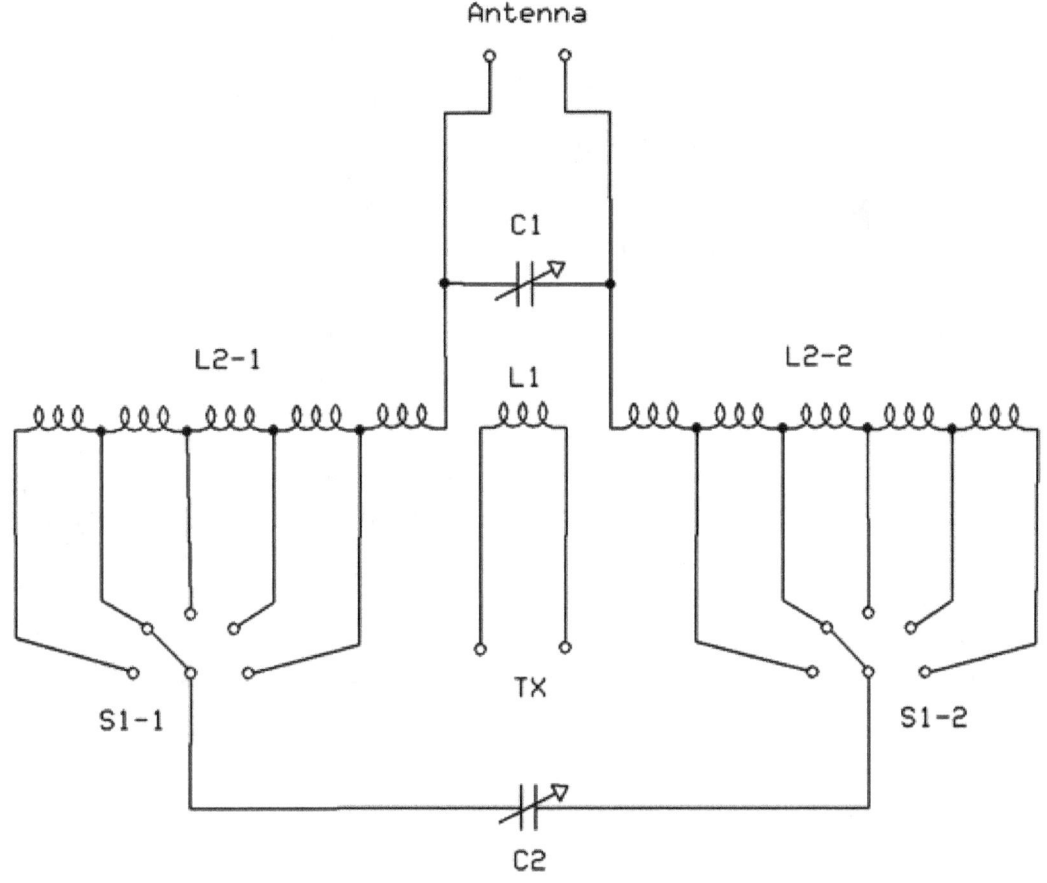

Figure 1 Schematic of the Symmetrical ATU

Taps are from 8, 12, 13 and 14- turns from ends of the inductors (see **Figure 2**). The five taps were enough to tune antenna at all amateur's bands (including WARC). Taps of the inductor L1 placed at the other side of the dielectric plate.

It was used a three – plate rotary switch at the ATU. It allowed used the ATU at high power – up to 500- Wtts. Take a look to the **Figure 2**. Taps for upper bands for L2-1 and L2- 2 are connected to one plate (near the inductor) of the rotary switch. Taps for lower bands for L2- 1 and L2- are connected to separate plates (one plate for one inductor) of the rotary switch.

Capacitor C1 is 500- pF air capacitor from old tube receiver. Capacitor C2 should be high quality high voltage capacitor with maxima 150- pF. For 150- Wtts going to the Symmetrical ATU it would be enough to install air capacitor with 2… 3 mm gap between plates. However when I used the Symmetrical ATU with such capacitor connected to my amplifier FL-2100 it is happened sparking between plates of the C2. Vacuum variable capacitor (as seen on **Figure 2**) installed in the ATU resolved the problem.

References

(1) Gierlach, W.: *Das DARC Antennenbuch*. DARC-Verlag, 1994

Figure 2 Design of the Symmetrical ATU

VHF- UHF Antennas

www.antentop.org

Three Element YAGI Antenna for 145- MHz with Square Reflector

By: *Yuriy Skutelis, RN3DEK*

The antenna provides good F/B ratio. Antenna has input impedance 50- Ohm that allows fed the antenna directly through 50- Ohm coaxial cable. It was reached by special form of the reflector. **Figure 1** shows design of the antenna. Reflector of the antenna made of aluminum wire in diameter 4- mm. Elements of the antenna made of aluminum tube in diameter 8- mm. Boom of the antenna made from dielectric stuff.

Figure 2 shows input impedance of the antenna. **Figure 3** shows SWR of the antenna. **Figure 4** shows DD of the antenna.

MMANA File for the Three Element Yagi Antenna for 145- MHz with Square Reflector may be loaded at: http: // www.antentop.org/019/square_019.htm

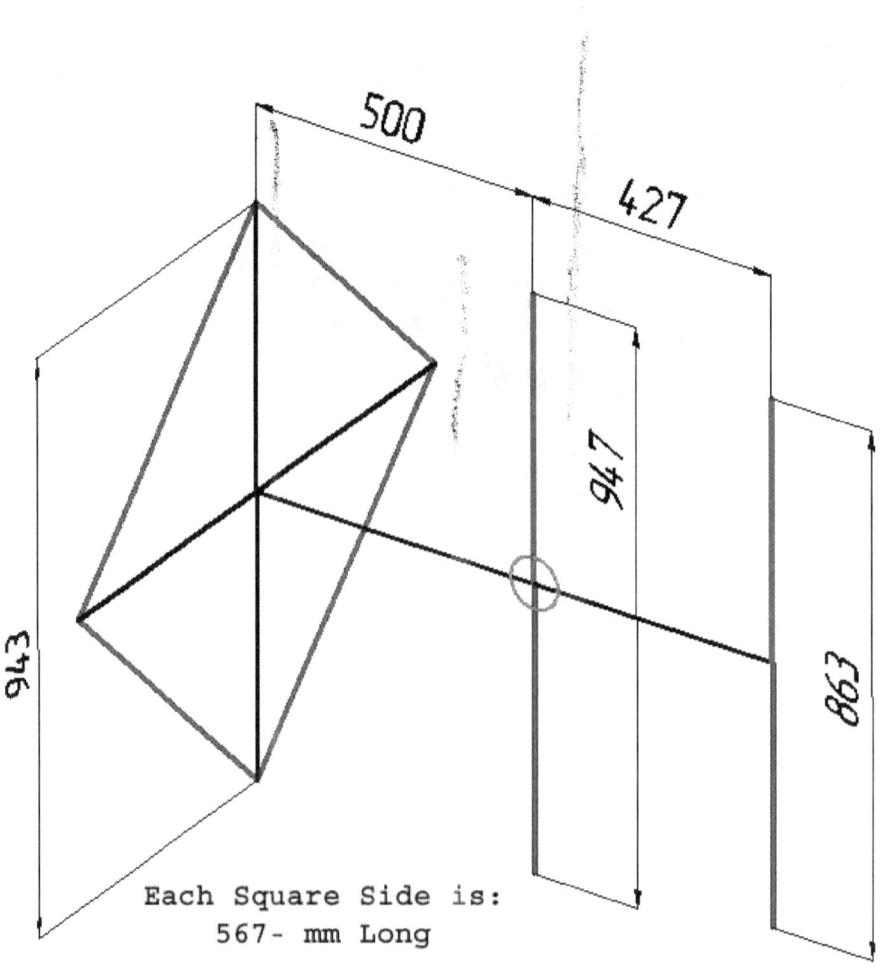

Figure 1 Three Element Yagi Antenna for 145- MHz with Square Reflector

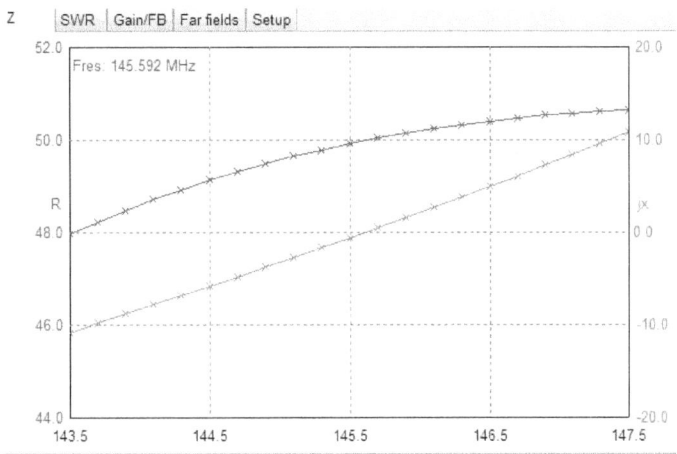

Figure 2 Z of the Three Element Yagi Antenna for 145- MHz with Rectangle Reflector

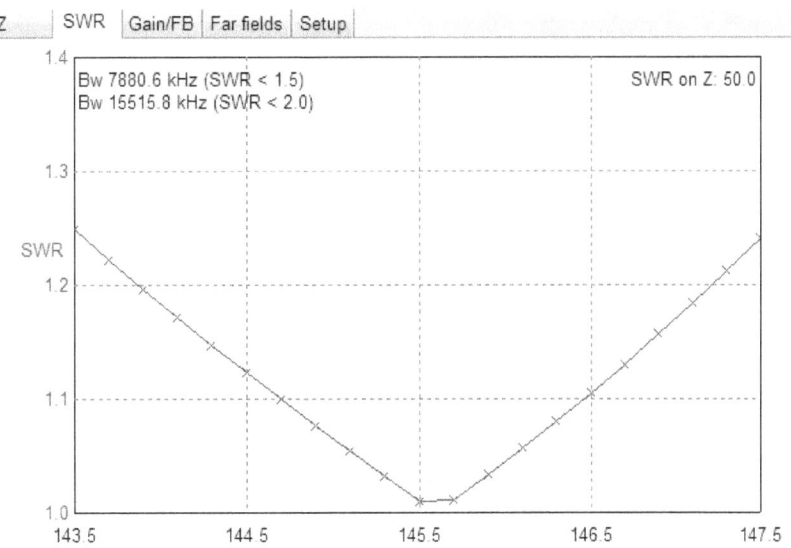

Figure 3 SWR of the Three Element Yagi Antenna for 145- MHz with Rectangle Reflector

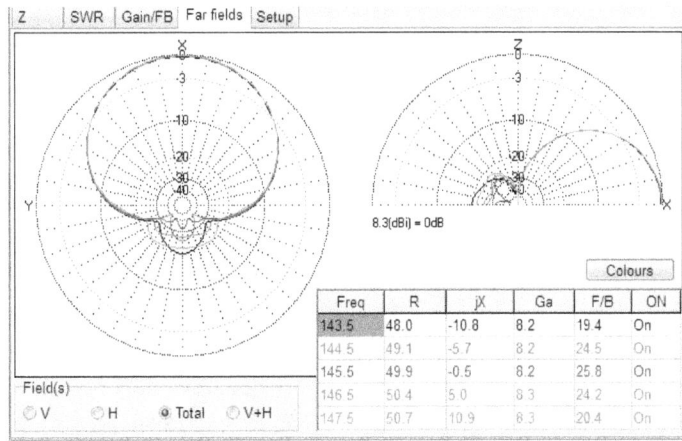

Figure 4 DD of the Three Element Yagi Antenna for 145- MHz with Rectangle Reflector

Three Element YAGI Antenna for 145- MHz with Rectangle Reflector

By: *Yuriy Skutelis, RN3DEK*

The antenna has F/B ratio at least 29 dB. It was reached by special form of the reflector. **Figure 1** shows design of the antenna. Reflector of the antenna made of aluminum wire in diameter 5- mm. Elements of the antenna made of aluminum tube in diameter 8- mm. Boom of the antenna made from dielectric stuff. Gamma matching used to feed the antenna. It is because "pure" antenna has input impedance close to 40- Ohm. **Figure 2** shows design of the gamma matching.

Figure 3 shows input impedance of the antenna. **Figure 4** shows SWR of the antenna. **Figure 5** shows DD of the antenna.

MMANA File for the Three Element Yagi Antenna for 145- MHz with Rectangle Reflector may be loaded at: http://www.antentop.org/019/rectangle_019.htm

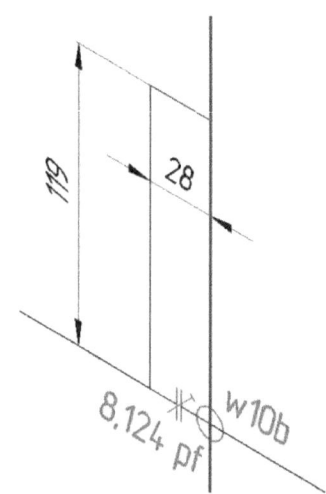

Figure 2 Gamma matching

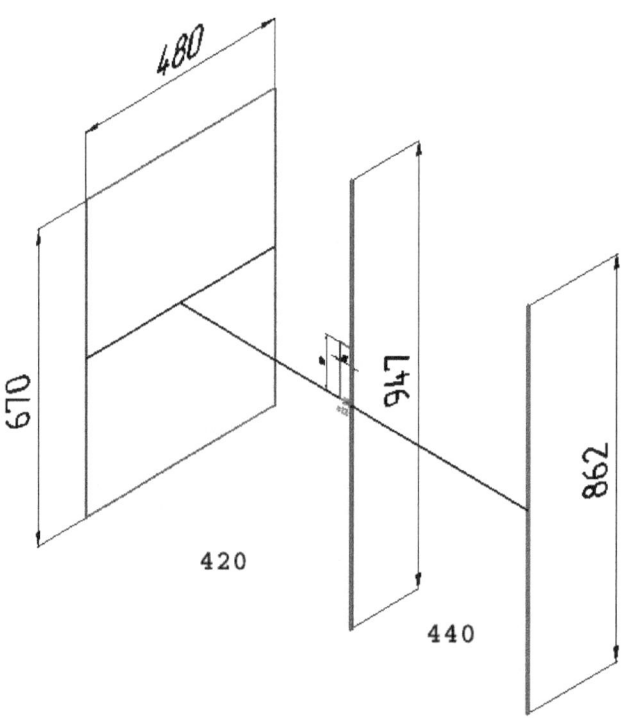

Figure 1 Three Element Yagi Antenna for 145- MHz with Rectangle Reflector

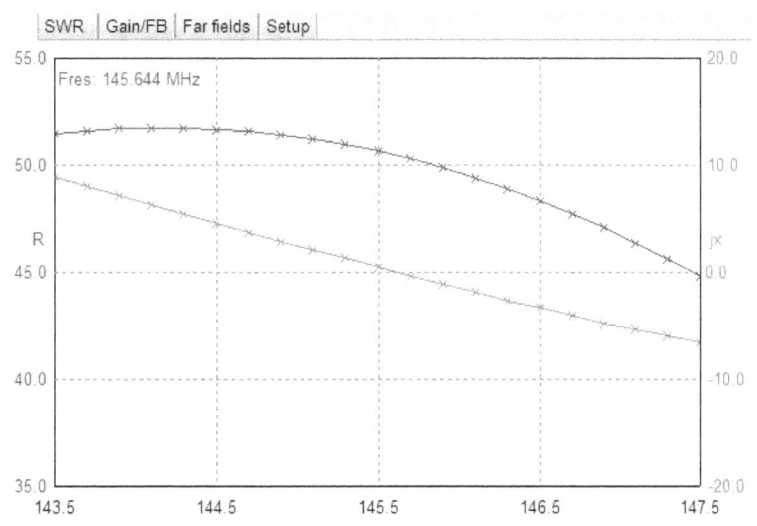

Figure 3 Z of the Three Element Yagi Antenna for 145- MHz with Rectangle Reflector

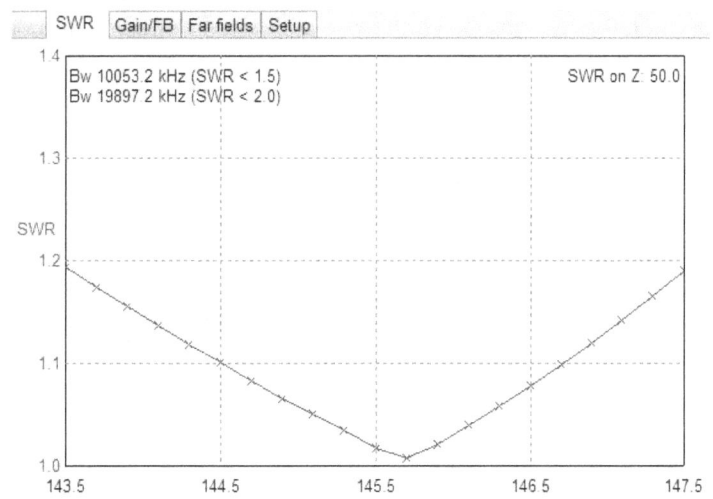

Figure 4 SWR of the Three Element Yagi Antenna for 145- MHz with Rectangle Reflector

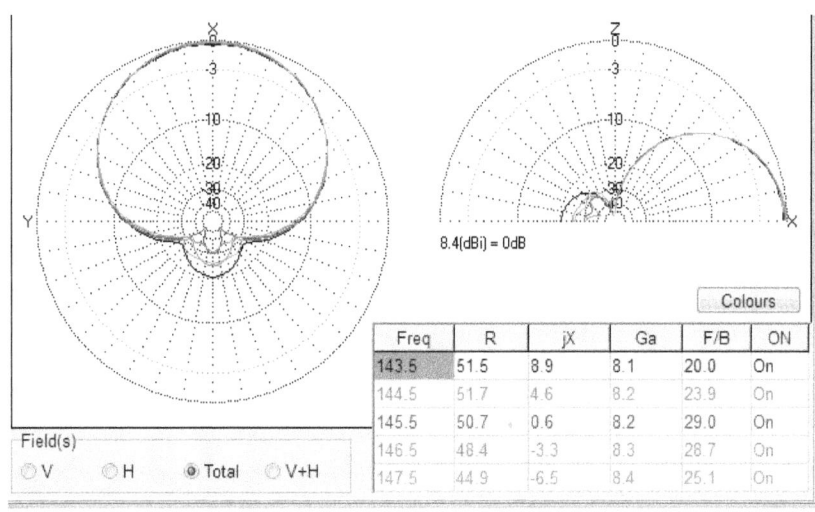

Figure 5 DD of the Three Element Yagi Antenna for 145- MHz with Rectangle Reflector

Four Element Antenna for Stack Design for 145- MHz Band

The publication is devoted to the memory UR0GT.

Credit Line: Forum from:
www.cqham.ru

By: Nikolay Kudryavchenko, UR0GT

The four element YAGI is designed for installing in Four Element Stack Antenna System. The antennas not critical to nearest objects. Four such antennas are installed at corners of a quad.

Two groups of antennas are connected through a 75- Ohm coaxial cable (the same length from each antenna) in bridge. It is 37.5- Ohm. Then the two groups connected to bridge each group through lambda/4 transformer from 50- Ohm coaxial cable. It is 33.3 Ohm.

The MMANA model of the Triangle Vertical Antenna for 20, 15 and 10- meter Bands may be loaded: http: //www.antentop.org/019/stack_ur0gt_019.htm

Such impedance could be easy matched with 50- Ohm coaxial cable. Sure it is possible use just one such antenna with your 145- MHz transceiver.

The antenna does not require a symmetrical unit at feed points. Boom for the antenna may be made as from metal either from dielectric (optimal design) stuff.

Figure 1 shows the four elements YAGI for stack design. **Figure 2** shows Z of the antenna. **Figure 3** shows SWR of the antenna. **Figure 4** shows DD of the antenna.

73! UR0GT

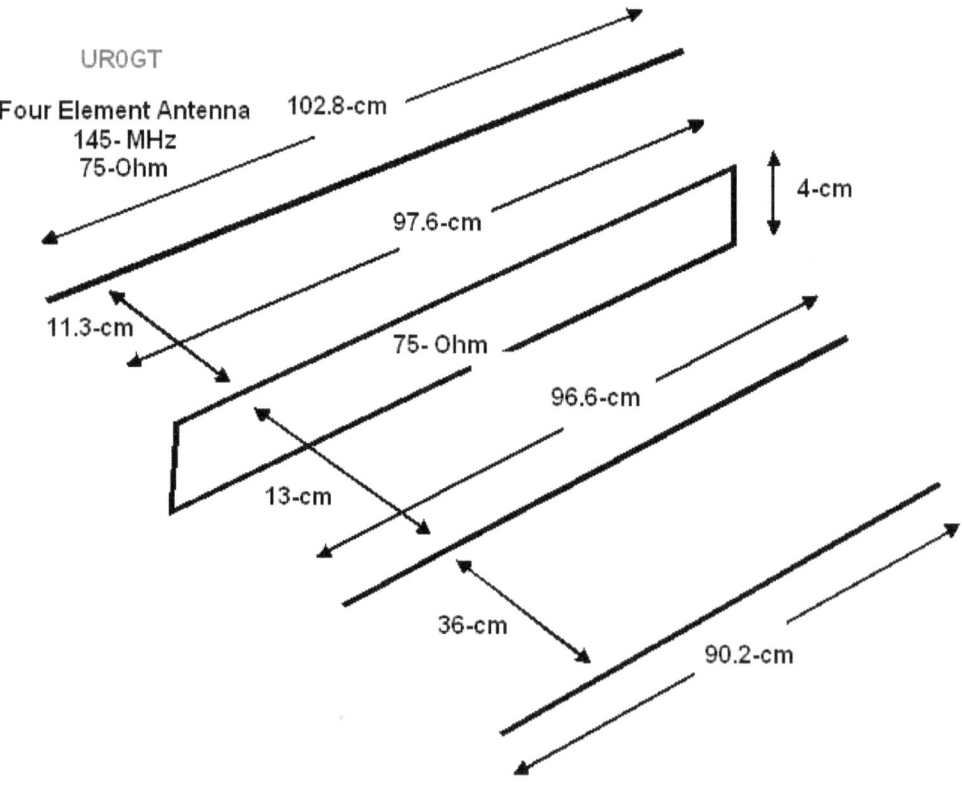

Figure 1 Four Elements YAGI for Stack Design

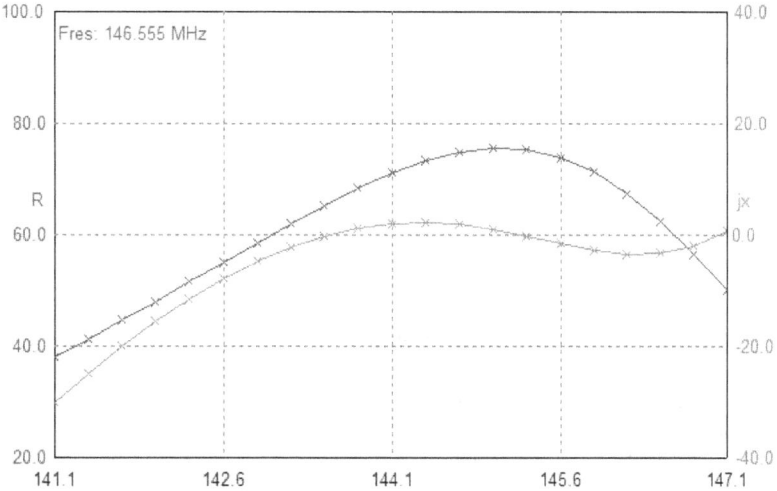

Figure 2 Z of the Four Elements YAGI for Stack Design

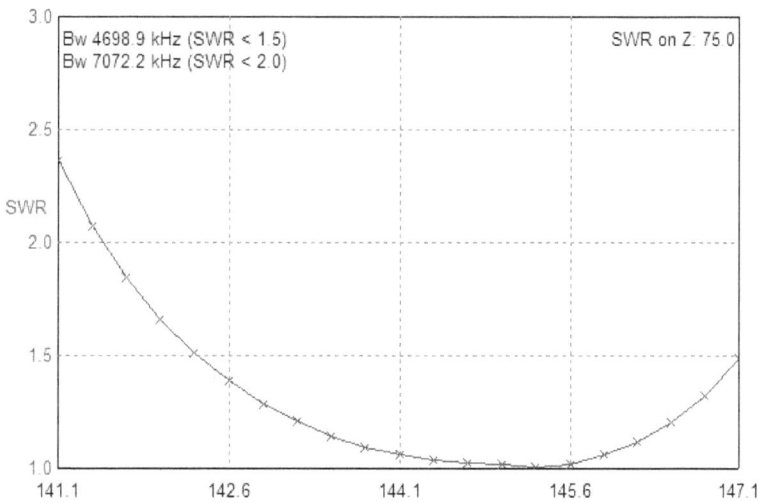

Figure 3 SWR of the Four Elements YAGI for Stack Design

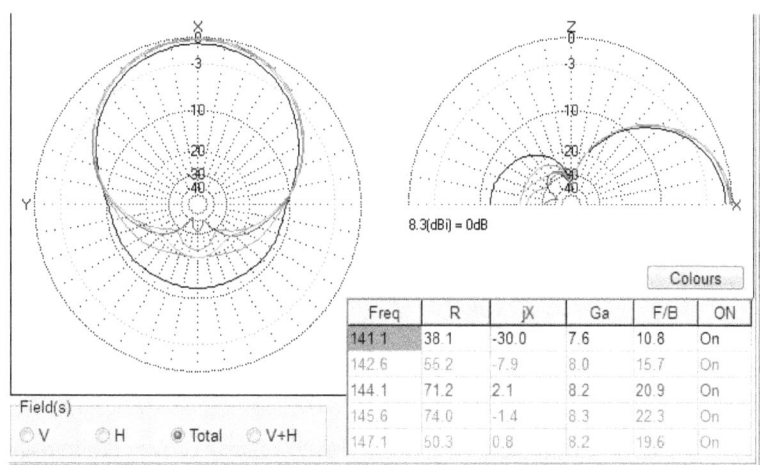

Figure 4 DD of the Four Elements YAGI for Stack Design

Vertical Antenna 5/8 Lambda for 70- cm Band

Credit Line: Forum from:
www.cqham.ru

By: Antonhax

At first the antenna was modeled by copper wire. Then the antenna was made on the base of bicycle wheel spokes. **Figure 1** shows design of the antenna. Inductor of the antenna is wound on a form in 8- mm diameter. After that we have got inductor with 9- mm ID. Central wire of the RF connector is glued by termo- glue for water protection.

Antenna is tuned to the resonance by shortening length of the vertical part of the antenna. Precise tuning may be done by chosen the tap from the inductor.

Figure 2 shows ready antenna. **Figure 3** shows tap from inductor of the antenna. **Figure 4** shows vertical at inductor of the antenna. **Figure 5** shows SWR Vertical Antenna 5/8 Lambda for 70- cm Band. **Figure 6** shows Z Vertical Antenna 5/8 Lambda for 70- cm Band

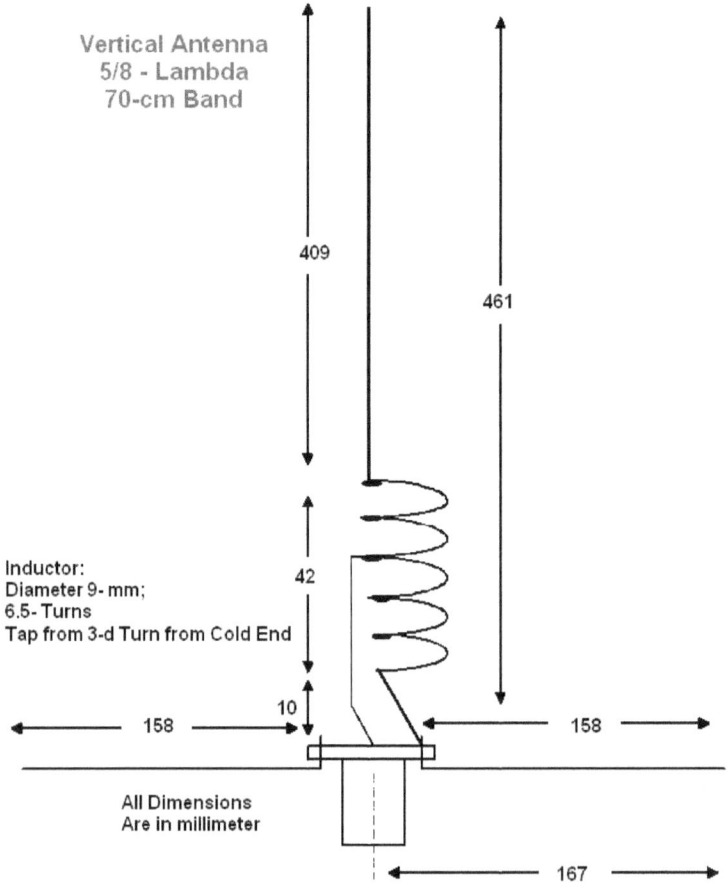

Figure 1 Design of the Antenna Made on the Base of Bicycle Wheel Spokes

Figure 2 Ready Antenna

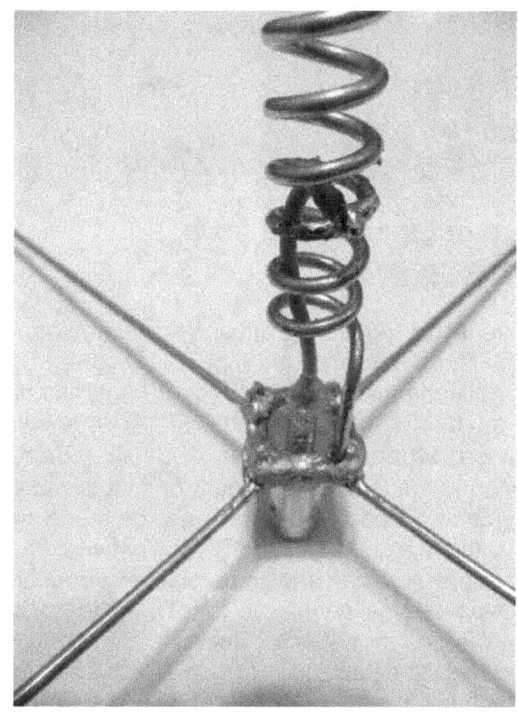

Figure 3 Tap from Inductor of the Antenna

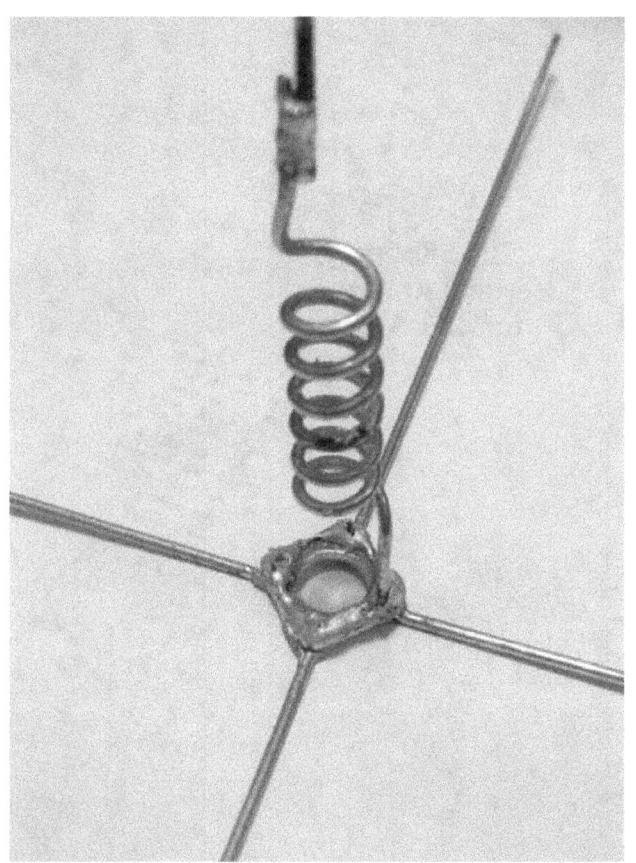

Figure 4 Vertical at Inductor of the Antenna

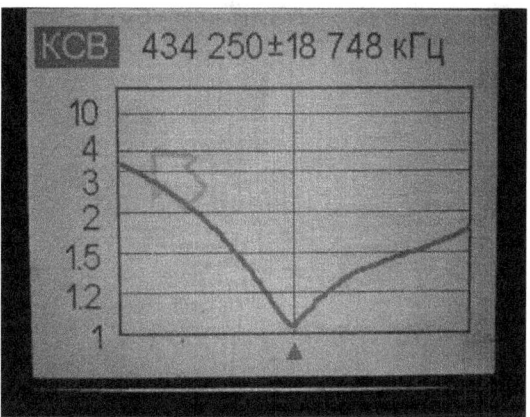

Figure 5 SWR Vertical Antenna 5/8 Lambda for 70- cm Band

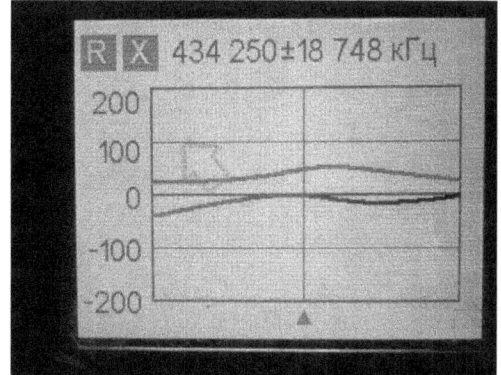

Figure 6 Z Vertical Antenna 5/8 Lambda for 70- cm Band

Broadband Vertical for 430- MHz Band

The publication is devoted to the memory UR0GT.

Credit Line: Forum from:
www.cqham.ru

By: Nikolay Kudryavchenko, UR0GT

Broadband Vertical Collinear Vertical antenna designed for 430- MHz Band. The antenna has Diagram Directivity with low-altitude maxima to the ground. Antenna has passband near 70- MHz at SWR 1.5:1.0. **Figure 1** shows design of the Broadband Vertical for 430- MHz Band.
Antenna made from aluminum or copper rod in 3- mm diameter. Capacitor 0.22- pF may be made structurally. Just thin dielectric- mika or Teflon between the two vertical vibrators. Antenna should be places at conductivity surface or have several counterpoises. The antenna has 200- Ohm input impedance. Such input impedance may be easy matched at one Band but there are some difficulties to match such impedance in wide band by simple methods.

The MMANA model of the Broadband Vertical for 430- MHz Band may be loaded: http://www.antentop.org/019/broadband_ur0gt_019.htm

Figure 2 shows input impedance of the antenna at 430- MHz Band. **Figure 3** shows SWR of the antenna at 430- MHz Band. **Figure 4** shows DD of the antenna at 430- MHz Band.

Figure 5 shows input impedance of the antenna at 100- MHz pass band at 433- MHz central frequency. **Figure 6** shows SWR of the antenna at 100- MHz pass band at 433- MHz central frequency. **Figure 7** shows DD of the antenna at 100- MHz pass band at 433- MHz central frequency.

73! de UR0GT

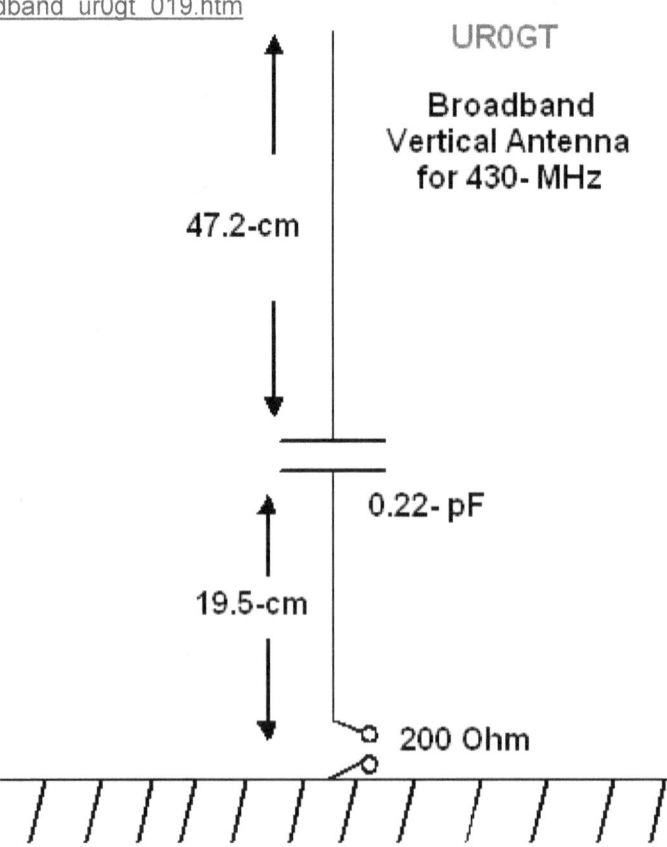

Figure 1 Design of the Broadband Vertical for 430- MHz Band

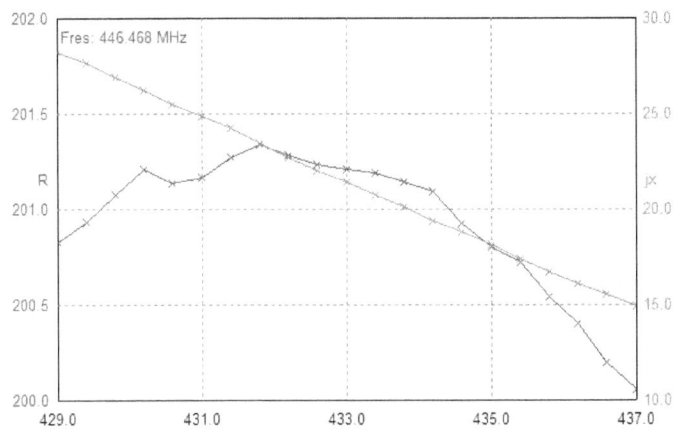

Figure 2 Z of the Antenna at 430- MHz Band

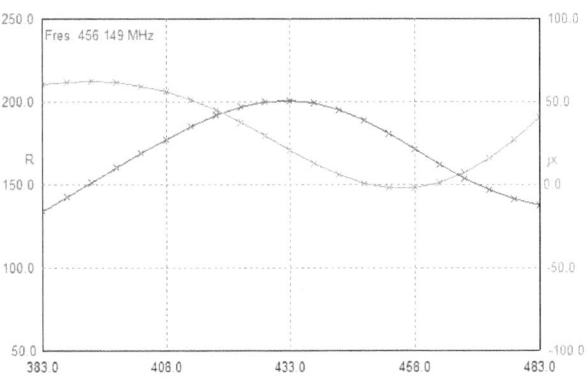

Figure 5 Z of the Antenna at 100- MHz Pass Band at 433- MHz Central Frequency

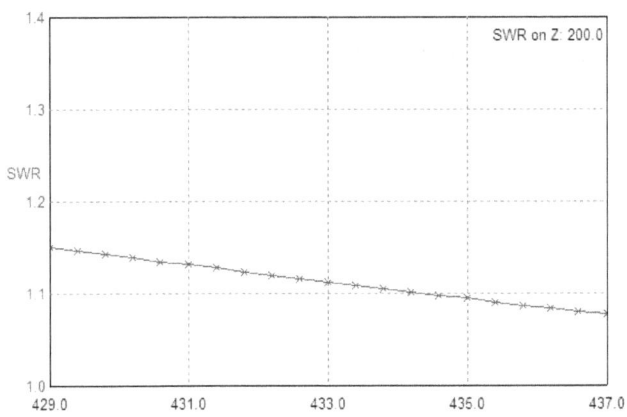

Figure 3 SWR of the Antenna at 430- MHz Band

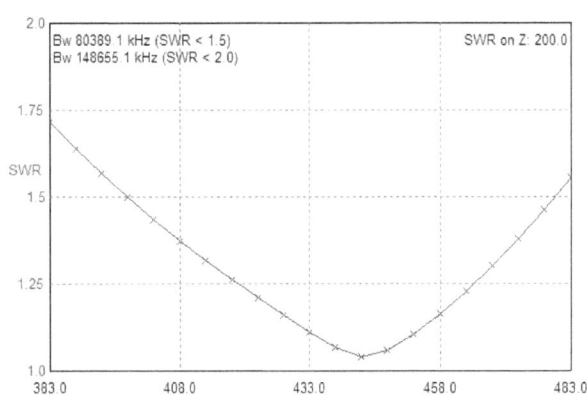

Figure 6 SWR of the Antenna at 100- MHz Pass Band at 433- MHz Central Frequency

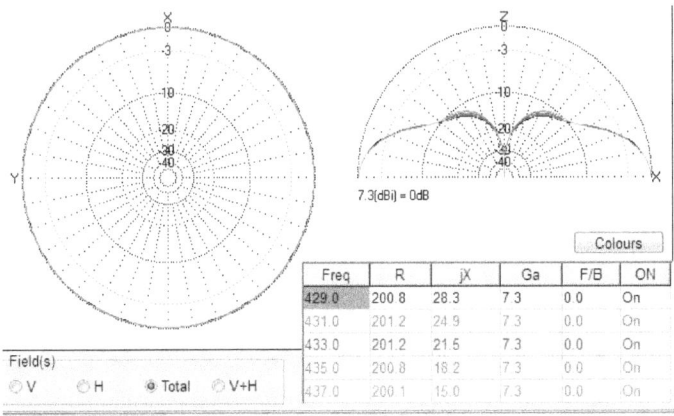

Figure 4 DD of the Antenna at 430- MHz Band

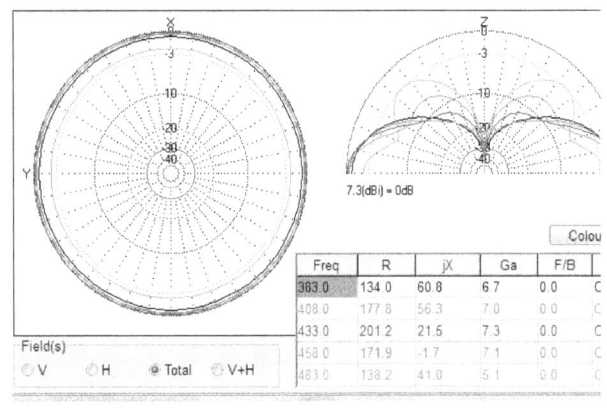

Figure 7 DD of the Antenna at 100- MHz Pass Band at 433- MHz Central Frequency

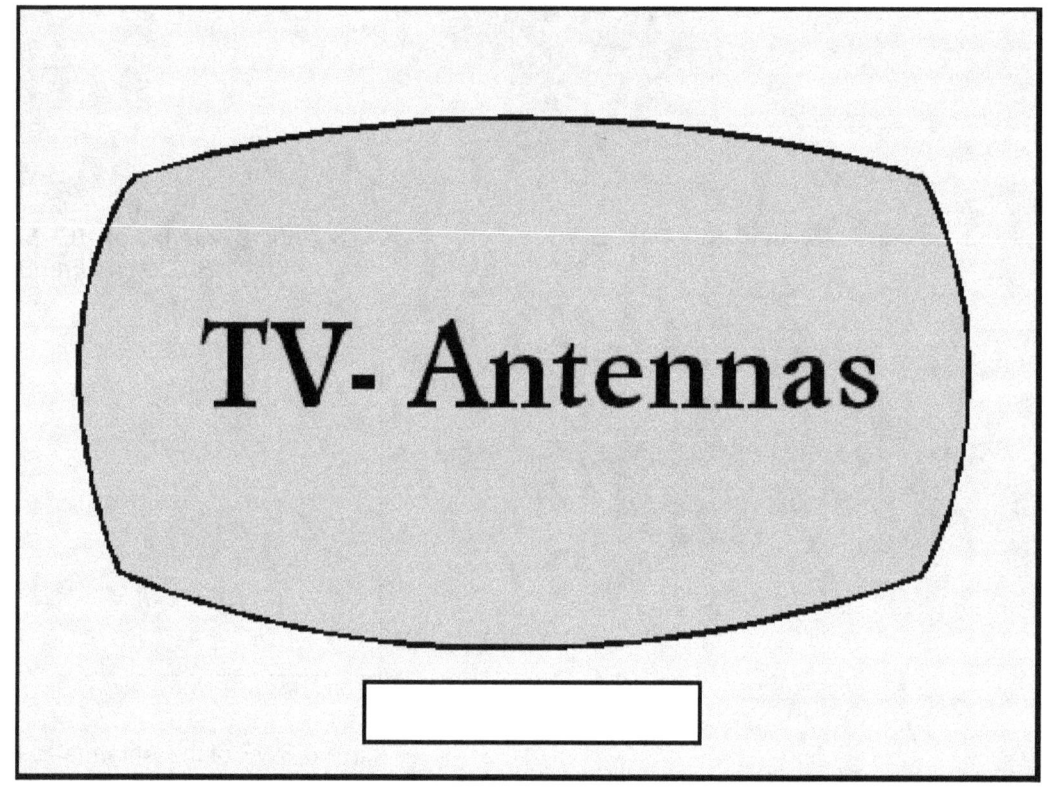

TV Antennas for Distance Receiving

By: Leonid Pozdnyakov
Credit Line: Radio # 10, 1953, pp.: 53- 54.

At the original articles published at Radio # 10, 1953, there were described several antennas for distance receiving TV broadcasting stations. Below it is described one of those antennas- it is a Rhombic Antenna. Rhombic Antenna is easy to make and at the same time has perfect parameters.

Rhombic Antennas are easy to build and at the same time has high gain and good diagram directivity. However the antennas have some lack. Such antennas required lots space for installations and need at least for masts instead one that used to support traditional directional antennas.

Figure 1 shows design of the Rhombic Antenna. Rhombic Antenna is a rhomb that hang up horizontally at the ground. Feeder is connected on to one sharp angle of the rhomb. Terminated resistor is connected on to far sharp angle of the rhomb. The resistor's value should be equal to the impedance of the rhomb at the working frequencies of the antenna. As usual the value is near 700- Ohm. Working frequencies of the antenna may have pass band in hundreds megahertz. So using such matched resistor allows create a super broadband antenna that has impedance near 700- Ohm at the frequencies window in several hundred megahertz.

High gain and high directivity of the rhomb antenna could be explained by combining gain and diagram directivity of the parts of the antenna. The antenna consists of four wires with traveling wave. **Figure 2** shows the combination. Each wire has own gain and diagram directivity.

Front Cover **Radio # 10, 1953**

The gain and diagram directivity depends on ratio the length of the wire to the working wavelength. So, the summary gain and diagram directivity depends on the ratio the length of the wire to the working wavelength and to the sharp angle of the rhomb.

АНТЕННЫ ДЛЯ „ДАЛЬНЕГО" ПРИЕМА
ТЕЛЕВИДЕНИЯ

Title of the Article

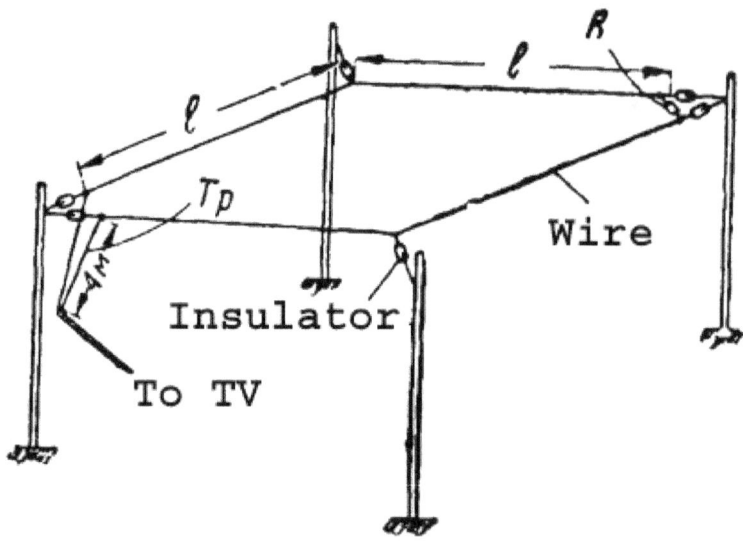

Figure 1 Rhombic Antenna

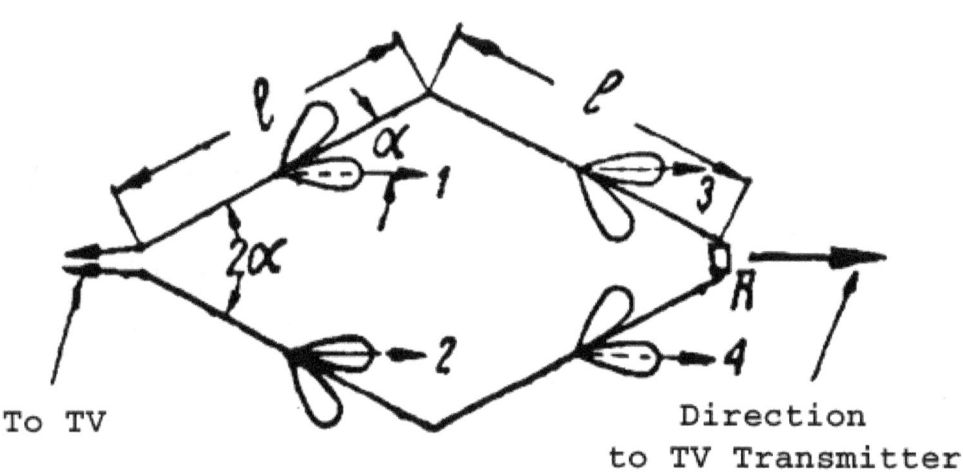

Figure 2 Rhombic antenna is Combination of Four Wires with Traveling Wave

Table 1 shows data for Rhombic Antenna with different parameters. To keep such parameters antenna should be placed above the ground at height not less the 2- 3 wavelength of the working band of the antenna.

Antenna may be fed by open ladder line with wave impedance 300… 600- Ohm. At this case the antenna could be matched at all working frequencies band. Antenna may be fed through a coaxial cable when two simple matching transformers are used. **Figure 3** shows feeding Rhombic Antenna through a coaxial cable. First transformer is a broadband transformer made on two wire ladder line. It is two wires line with varying wave impedance on the length.

The wave impedance of the line changes from 700- Ohm at rhomb side to 300- Ohm at coaxial cable side. At coaxial cable side the coaxial cable should not connect straight away to the line. Coaxial cable connected to the line through a symmetrical transformer 4:1 made on lengths of the used coaxial cable. The transformer makes symmetrical and provides matching of the Rhombic Antenna to coaxial cable. Loop of the coaxial cable should have electrical length lambda/2. To calculate such transformer you need to know the shortening coefficient of the used coaxial cable.

Table 1 Data for Rhombic Antenna with Different Parameters

Length Side of the Rhomb L in Lambda	Sharp Angle of the Rhomb 2 alpha Degree	Gain of the Antenna	Directivity Gain
2	90	4	13
3	70	7	20
4	60	9	25
5	50	13	36
7	40	20	52

It is possible to use coaxial cable with any wave impedance- 50 or 75 Ohm. Matching impedance of the 4:1 transformer depends on the coaxial cable. At 50-Ohm cable it is got transformer 200:50- Ohm, at 75-Ohm cable it is got transformer 300:75- Ohm. Transformer 300:75- Ohm should have best matching result with open line transformer. There are lots link in the internet how the transformer may be calculated. One of them is: http://www.n-lemma.com/calcs/dipole/balun.htm. When the coaxial cable symmetrical transformer is used the broadband of the antenna depends on the broadband of the transformer. As usual coaxial cable transformer has good matching at the 5% frequencies band calculated from the central working frequency of the transformer. So, when such transformer is used the broad band of the Rhombic Antenna is limited to pass band of the transformer.

Antenna may be made from a strand wire in diameter 2… 3- mm. It may be copper, aluminum or bimetal (with copper or aluminum layer) wire. Terminated resistor at the antenna may be any small power non-inductive resistor. This one should be protected from atmospheric influences.

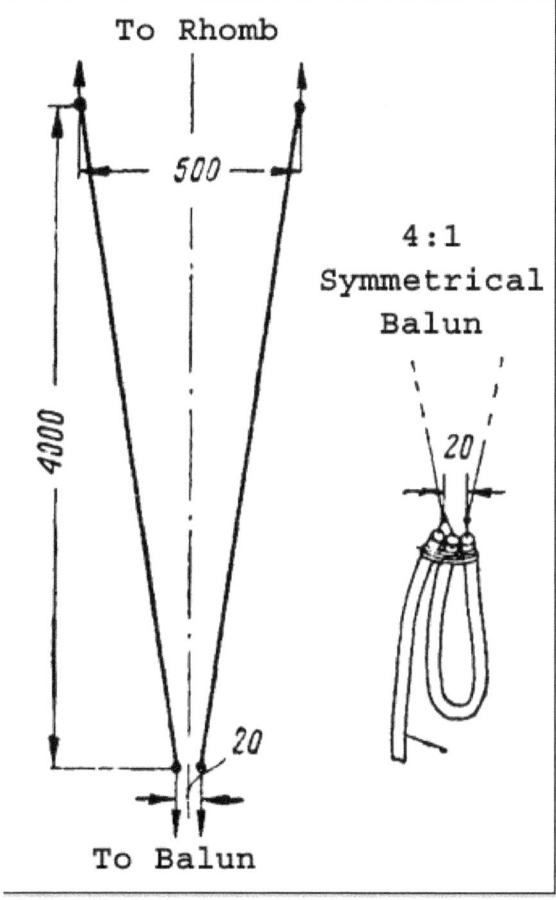

Figure 3 Feeding Rhombic Antenna through Coaxial Cable

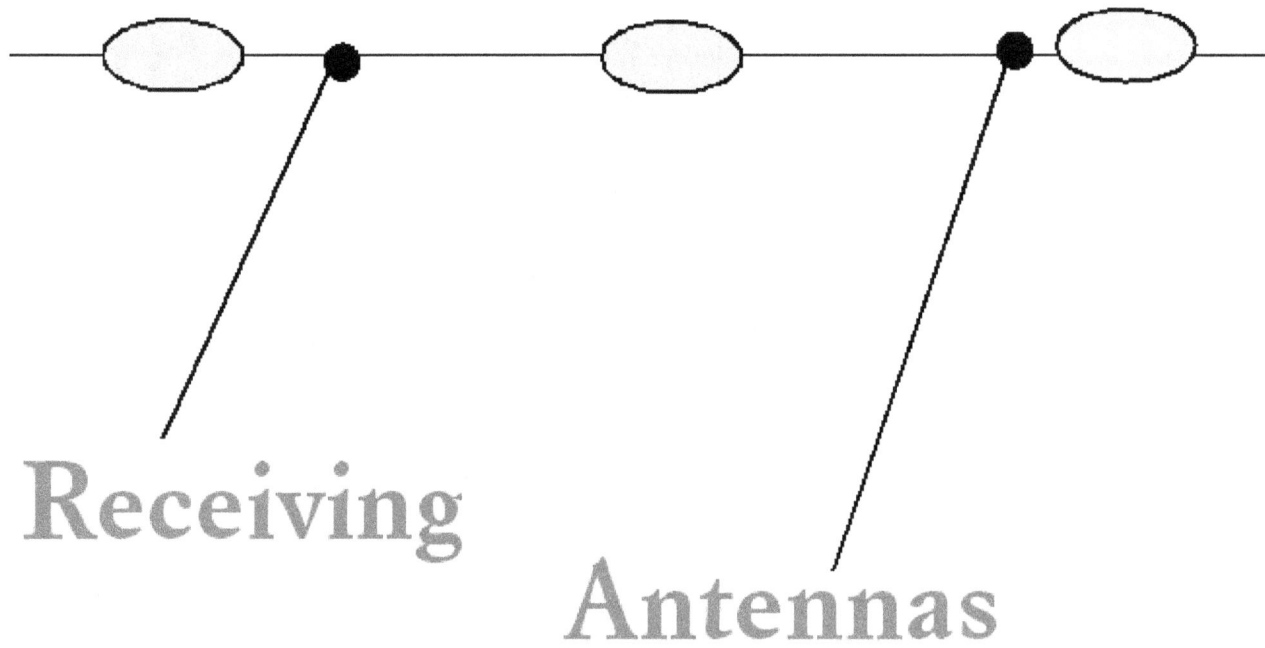

UB5UG Horizontal Receiving Antenna

By: Yuri Medinets, UB5UG
Credit Line: Radio # 12, 1977. P. 25

At modern city to use a separate receiving antenna may be only one variant to be on the Air. Interferences from nearest electronics devices could force it. Below here it is described receiving antenna from far 70-s that may solve the hard modern situation.

The antenna works at 80-, 40-, 20-, 15- and 10- meter Bands. Antenna receives mostly radio waves having horizontal polarisation. Industrial electrical interferences as usual has vertical polarisation. Commercial and ham radio- stations as rule have antennas with vertical polarisation. So, the antenna allows decrease at receiving input both as interferences and signals from nearest radio- stations. At right installation the antenna reduces unwanted signals with vertical polarization up to 20- 30- dB. However the antenna would effectively work for receiving DX station because the stations have as vertical as horizontal components.

Antenna has circle diagram directivity in horizontal plane and 8- shape diagram directivity in vertical plane. It is horizontal loop antenna that feed by transmission line in standing wave mode. **Figure 1** shows design of the antenna. Loop antenna made from two length of the same coaxial cable (item P/2). Feeder L made from the same coaxial cable as loop of the antenna. It is possible use any coaxial cable- 50 or 75- Ohm.

Antenna has electrical length lambda/2 at 80- meter Band, lambda at 40- meter Band, two lambda at 20- meter Band, three lambda at 15- meter band and four lambda at 10- meter Band. Receiving part of the antenna is only loop. When the loop made symmetrically relative to the axis AB the part L cannot participate in receiving. Antenna has input impedance in several Ohms. The antenna may be matched with receiver by some known methods. Antenna has efficiency in several percent at 80- meter Band to several tens at 10- meter band. It is possible to increase the efficiency of the antenna at low bands by increasing of the diameter of the antenna loop.

Front Cover Radio # 12, 1977

ГОРИЗОНТАЛЬНАЯ ПРИЕМНАЯ АНТЕННА

Title of the Article

However, the antenna lost circular diagram directivity at P more the 0.35 lambda.

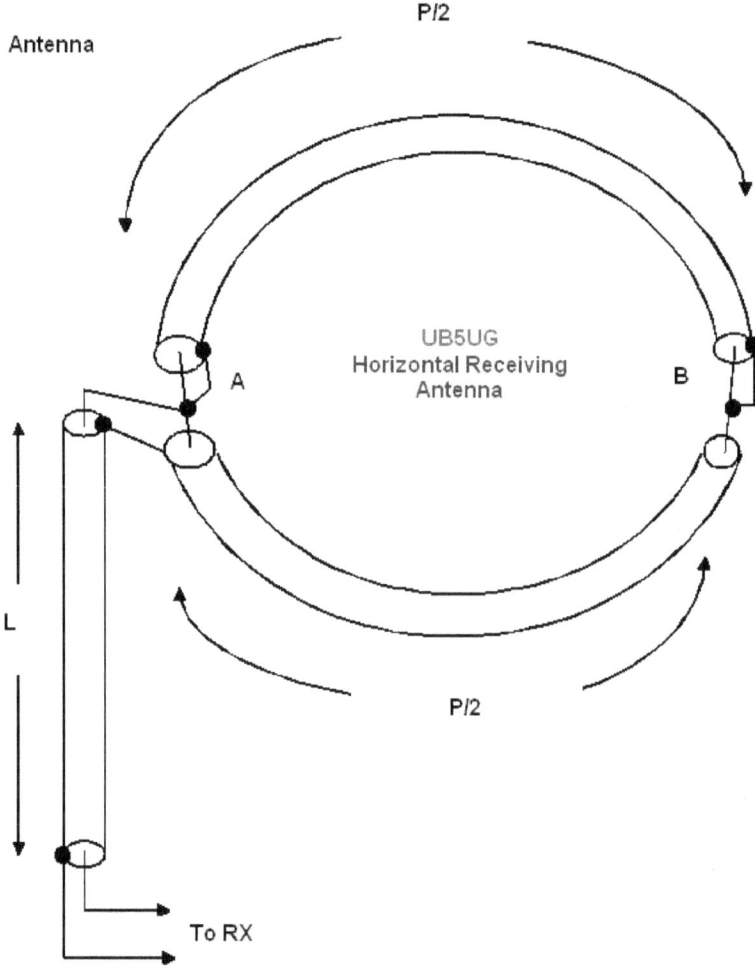

Figure 1 Horizontal Receiving Antenna

Take in attention the limitation (P less the 0.35 lambda), and count desirable quantity of the bands (80, 40, 20, 15 and 10) and take that electrical length of the antenna at lowest band is lambda/2 it is possible to find dimensions for the parts of the antenna according below formula.

$$\frac{P}{2} + \left(\frac{P}{2} + l\right) K_y = \frac{\lambda_{max}}{2}$$

There is at the formula "Ky" that is shortening coefficient for coaxial cable. As usual the shortening coefficient for coaxial cable with polyethylene dielectric is equal to 1.52, and shortening coefficient for coaxial cable with teflon dielectric is equal to 1.44. Formula below helps find Length L.

$$l = \frac{\lambda_{max} - P(K_y + 1)}{2K_y}$$

It is going from the formula that optimal sizes for 5-Bands (80, 40, 20, 15, 10- meter Bands) antenna is: P= 4 meter, L= 24.4- meter.

Loop of the antenna should be done symmetrically relative to the axis AB. Loop may have any shape- circular, oval, square or rectangle. Loop may be placed on any form- space out on a wooden crest, lay on the dielectric base and so on. If loop is installed above a conductive surface it should be installed at height 1…- 2- meter above. Feeder should go athwart to the loop. It is very desirable that conductive subjects do not place near the loop on at least 2 meter distance. The subjects may destroy symmetrical of the antenna.

Insulated RX Transformer

By: Igor Grigorov, va3znw

At my shack I have used a Coaxial Antenna Switch Protax CSR- 5G to change devices switched to my antenna. **Figure 1** shows commutation graph and schematic of the switch. **Figure 2** shows the switch sitting at my table. The switch is very convenient for amateur operation in the Air. I can easy switch antenna from one transceiver (ICOM- 718) to another one (K1) or turn antenna to general coverage receiver (Hallicrafters S-85). I use the receiver to check propagation in the Air and just to catch some interesting HF- stations.

I have bought the switch at some Hamfest for 5 dollars. It is very reliable switch that worked at me without any problem. Somedays I discovered ads of the switch at old 73- Magazine from 1967. **Figure 3** shows old advertising of the switch from 73- Magazine # 1, 1967.

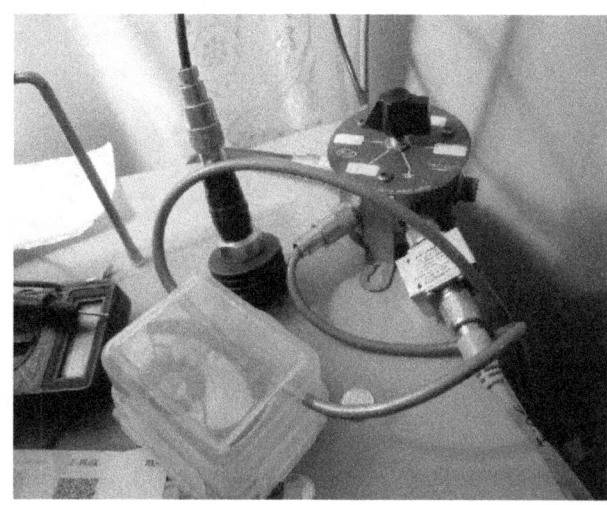

Figure 2 Protax Coaxial Antenna Switch Sitting at my Table

The switch costs $ 12.50 in 1967. I used online calculator (http://www.dollartimes.com/index.htm) to find value of the one dollar from 1967 to dollar in 2015.

It occurred that one dollar from 1967 cost 7.13 dollars from 2015. So the switch would cost 89 dollars if being sold in the 2015 year.

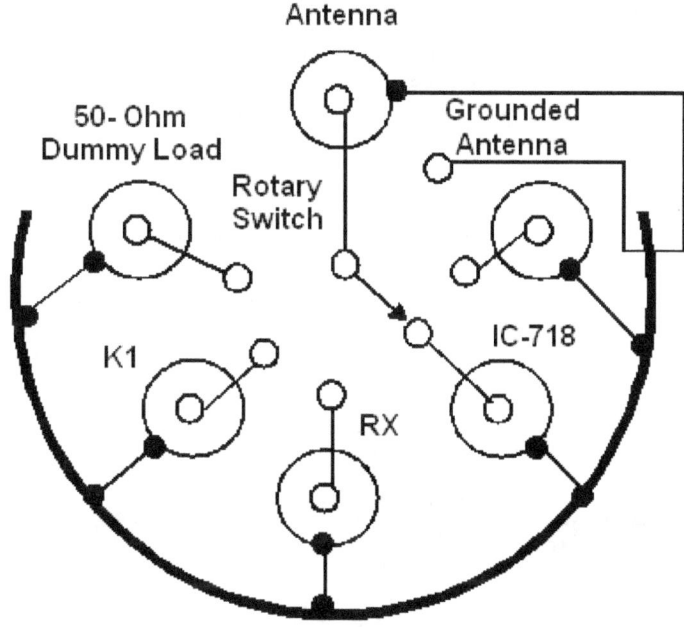

Figure 1 Commutation Graph and Schematic of the Switch

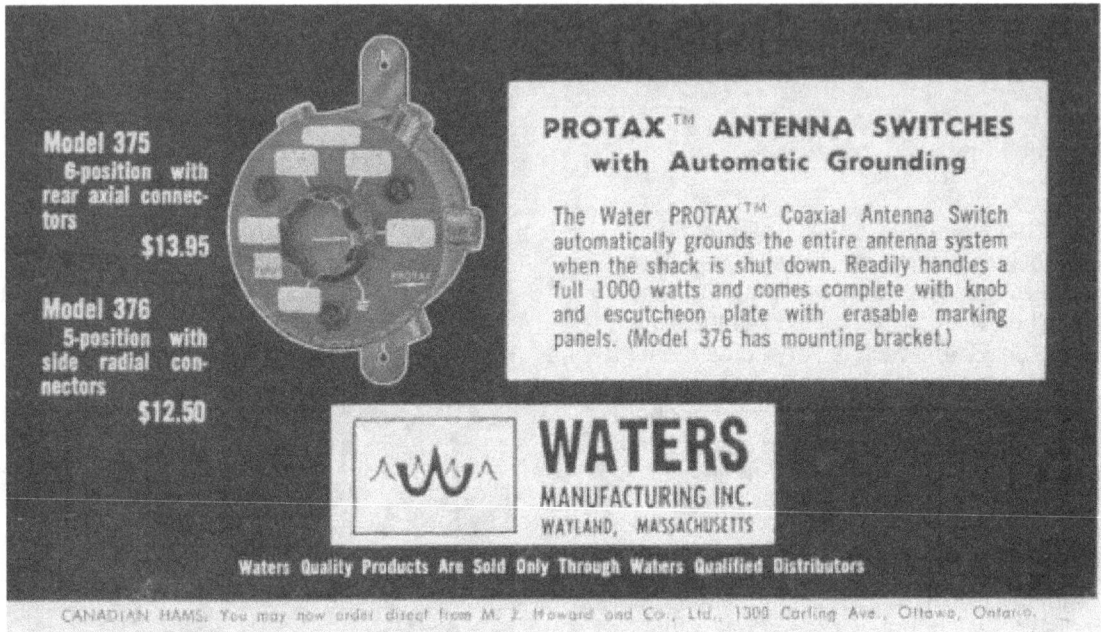

Figure 3 Advertising of the Coaxial Antenna Switch Protax from 73- Magazine # 1, 1967

Please, take attention that a 50- Ohm dummy load is switched there. Why? The switch has position when the antenna is grounded. It is very useful at lighting time or time when static could copy on to antenna wires. It is time of strong winds, snowfall or before lighting period. When antenna is grounded the static goes to the ground and could be not damaged devices that are switched to the Coaxial Switch. However it is worked well when antenna does not contains a ferrite transformer (s) in the design. If antenna had ferrite transformer in the design and the antenna is grounded to the natural ground, antenna current caused by static or lighting discharge may be strong enough to destroy the ferrite transformer.

I use to the switch with Beverage Antenna that is described at that issue (Antentop 01- 2015, pp.: 35- 41). My Beverage Antenna has ferrite matching transformer at the end. So theoretically the transformer may be damaged when switch grounded the antenna directly to the ground.

At professional radio communication an antenna that has ferrite parts in the design newer being directly grounded to the natural ground. Such antenna is grounded through a low ohm resistor that as usual has nominal equal to surge impedance of the transmission line. In this case the current going through the antenna ferrite parts would be limited by the resistor and the ferrite parts would not be destroyed by the lighting and static electricity. I use 50- Ohm coaxial cable so I turn on my antenna to a Dummy Load 50- Ohm when the antenna is not in use.

The main lack of the Coaxial Antenna Switch Protax CSR- 5G is that one position rotary switch is used at the device. So grounds of all devices connected together at the switch. It may be caused that interferences catching by one device may penetrate to another one having common ground. Professional Antenna Commutator as usual has 2 position rotary switch that commutate simultaneously input and ground of antenna or used device.

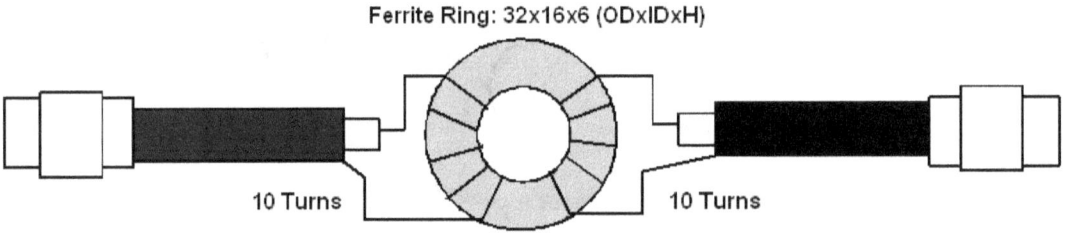

Figure 4 Design of the Insulation Transformer

In my case I mentioned that connecting on receiver Hallicrafters S-85 to the Coaxial Switch went to increasing noise at transceivers connected to the Switch. Insulation transformer helped me to remove the effect. **Figure 4** shows design of the transformer. **Figure 5** shows Hallicrafters S-85 on the bench.

Transformer was made with ferrite ring that I saved from an old Computer PIII. The ring had dimensions 32x16x6 (ODxIDxH). Permeability of the ring was unknown for me. Transformer had two identical inductors, each contained 10 turns coiled by wire in 0.5- mm diameter (24 AWG). The inductors were placed at opposite sides of the ferrite ring. Two lengths of coaxial cable in 20 cm were connected to the ring. The design was placed in plastic food box.

The transformer was installed between Coaxial Antenna Switch Protax CSR- 5G and coaxial cable going to receiver Hallicrafters S-85. **Figure 6** shows the transformer inside opened food box. . **Figure 7** shows the closed food box with insulation transformer. **Figure 2** shows the closed food box sitting beside the Coaxial Antenna Switch.

The insulation transformer allows use the Hallicrafters S-85 without any noise influences to transceivers connected to the switch. Reception of the receiver was improved with the insulation transformer. Stations sound clearly with the transformer. I was mention that there were less noise and clicks in the receiver especially at lighting and pre storm period.

Figure 5 Hallicrafters S-85

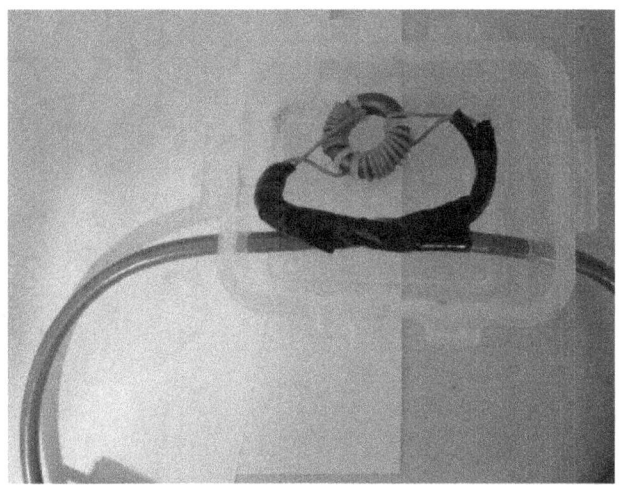

Figure 6 Insulation Transformer Inside Opened Food Box

I measured SWR when the transformer was connected to the 50- Ohm coaxial cable going from Beverage antenna and to MFJ-259B. SWR was in limit 3…4 up to 4- MHz. Then SWR was sharply increased and reach 25 at 20- MHz. However, such high SWR does not hinder the reception. It is more important for the receiver is to cut low frequencies interferences going from static, lighting and some home and industrial equipment. The insulation transformer does it in good way.

73! I.G., va3znw

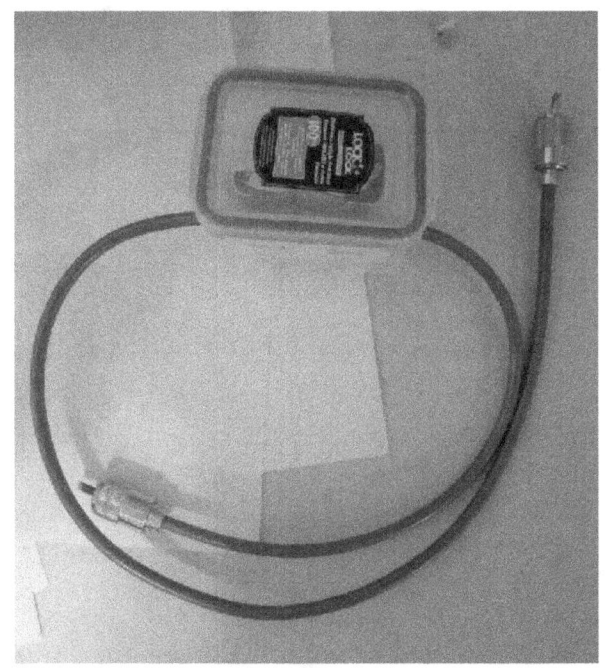

Figure 7 Closed Food Box with Insulation Transformer

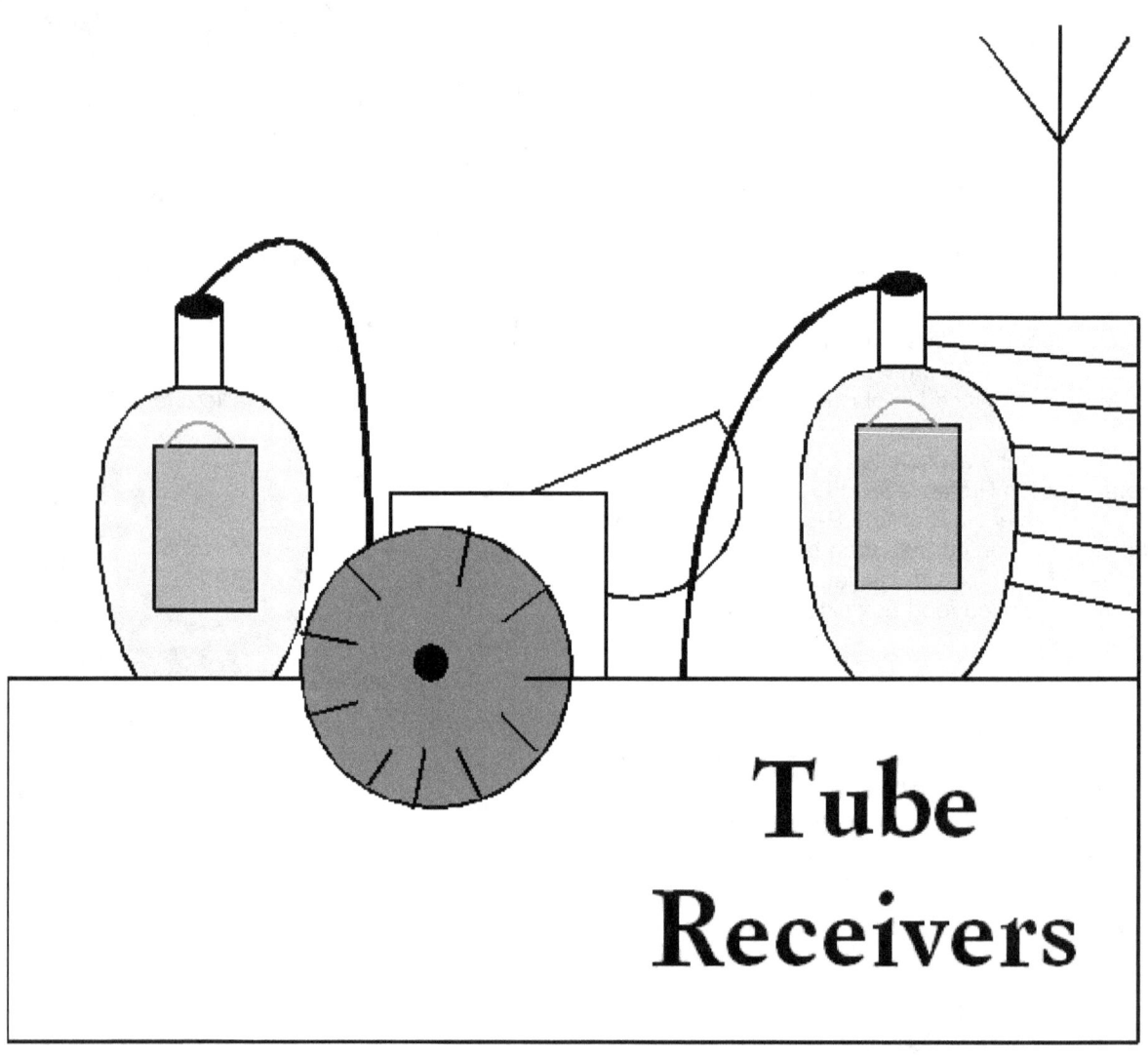

Simple Tube DC SSB Receiver

Credit Line: Forum from:
www.cqham.ru

http://www.cqham.ru/forum/showthread.php?15497-Простые-ламповые-конструкции/page4

By: R3KCR, Voronezh, Russia

It is simple experimental receiver that was made on two tubes- twin triodes. The receiver was tuned to 80- meter amateur band. Schematic of the receiver is shown on **Figure 1**.

Video of the receiver is:
https://yadi.sk/i/BRXlyqr8evRf8
Audio of the receiver is:
https://www.evernote.com/shard/s556/...0%BE%D1%81.amr

It is usual DC Receiver with phase suppression of the second side band. Book "Direct Conversation Technique for Radio Amateurs" (Antentop 01- 2015, p. 88) gives information how to tune and adjust such receiver. **Figure 2** shows design of the receiver. Receiver consumes current 4.3-mA. VFO consumes current 2 mA. **Figure 3** shows the current. **Figure 4** shows shape of the RF signal of the VFO. Receiver could work fine with antenna in several meter length. It is possible use players low – ohmic headphones turned on through transformer or high-ohmic (2x 2000-Ohm) phones turned on to tube plate through capacitor 47.0-microF/16-V.

Figure 2 Design of the SSB Receiver

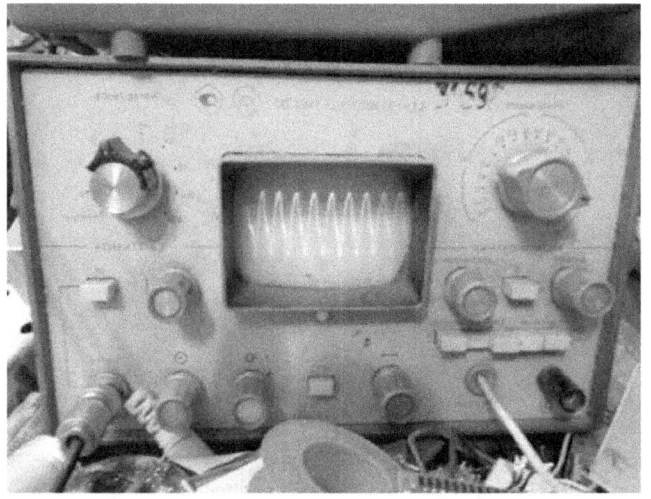

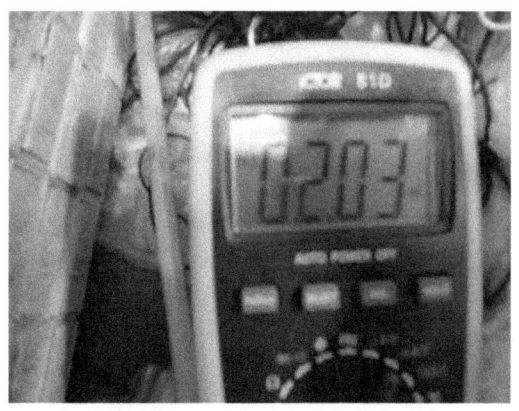

Figure 4 Shape of the RF Signal of the VFO

Figure 3 VFO Current

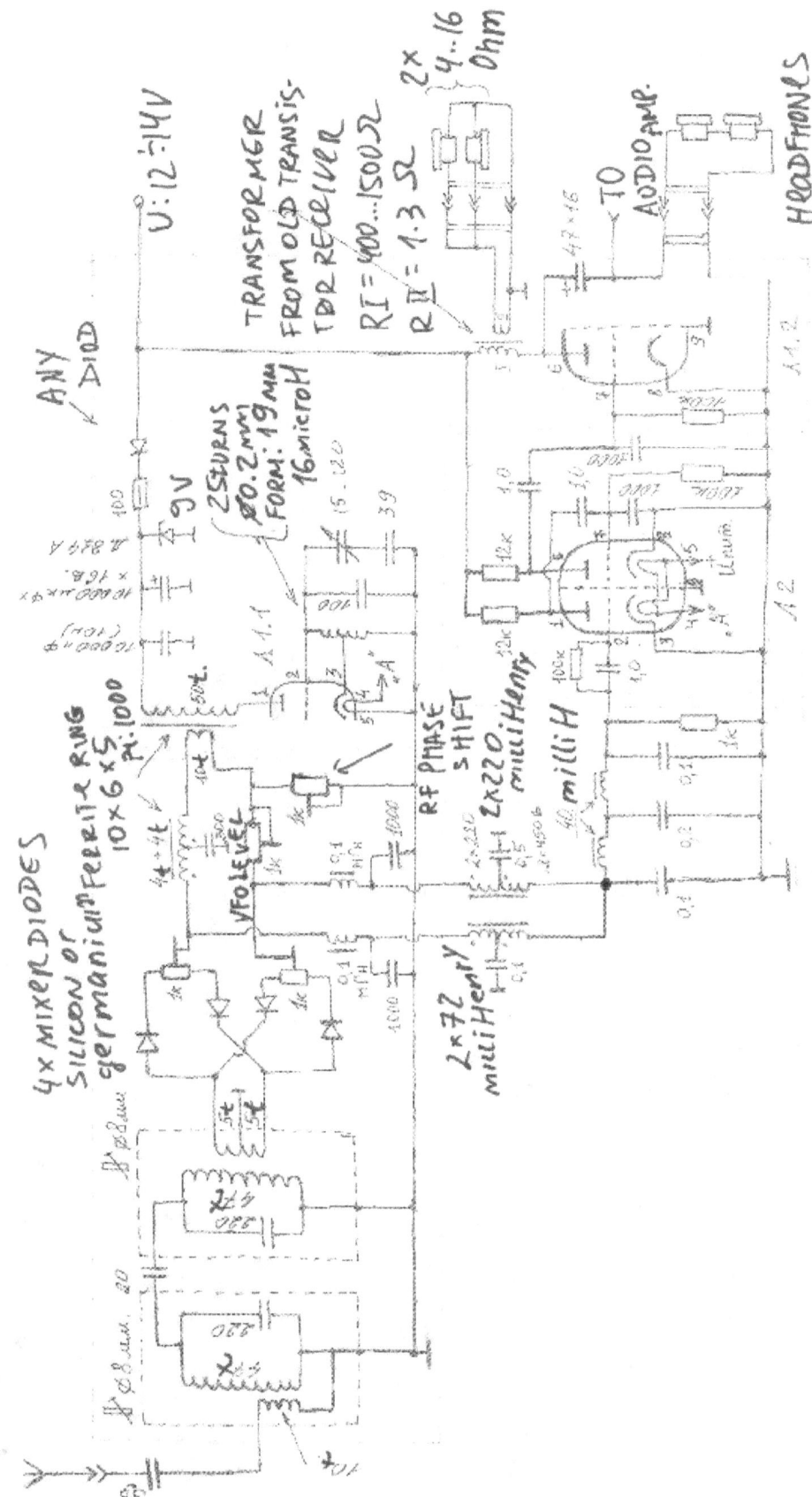

Figure 1. Schematic of Tube DC SSB Receiver

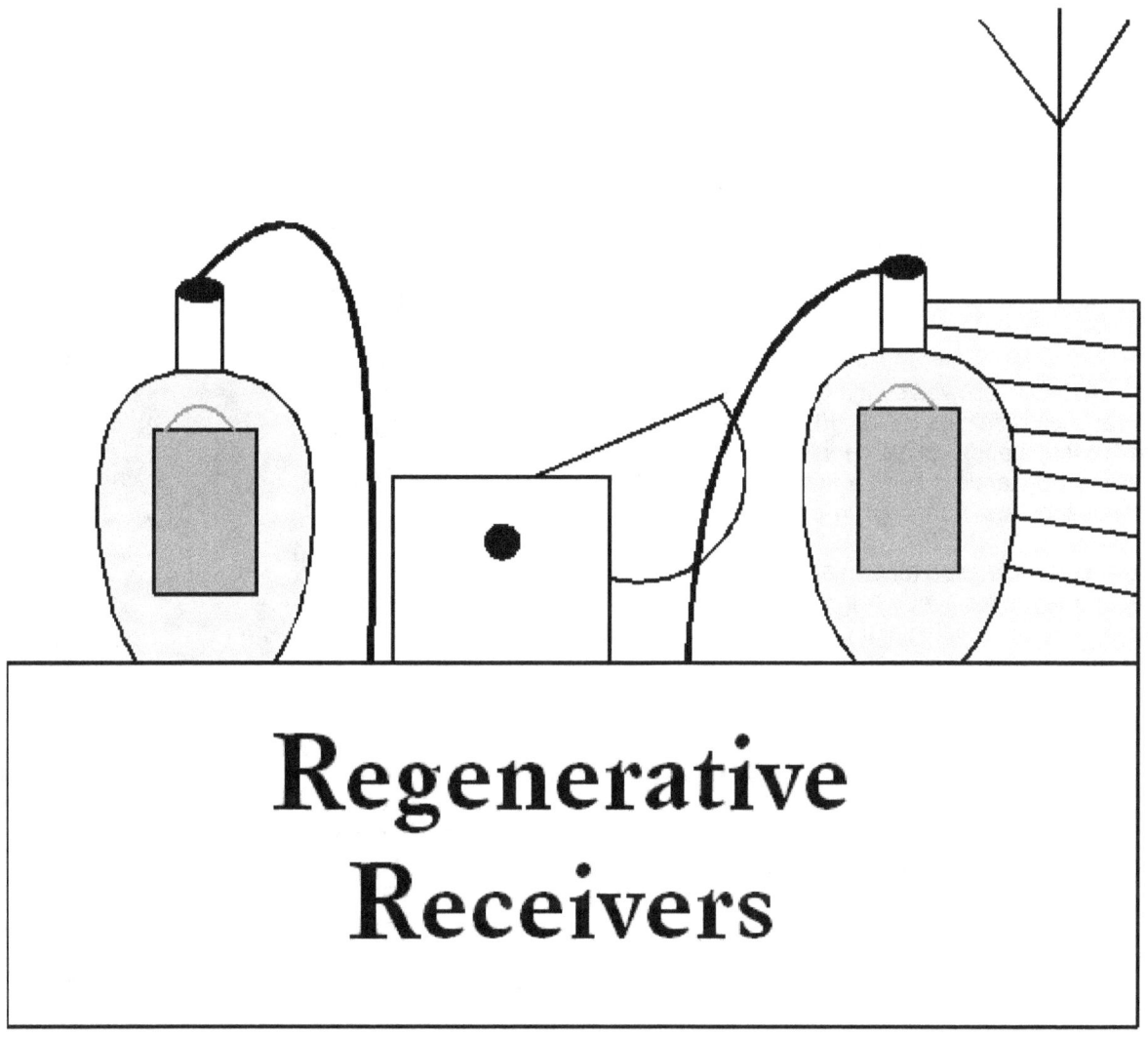

HF Receiver for Beginner Ham

Credit Line: Radio # 4, 1953: pp.: 26- 29

By: Viktor Lomanovich, UA3DH

The HF Tube Regenerative Receiver is a classical design of the Tube Era. Somebody gave me magazine with the article at the 70s. I did the receiver and got very good result. I have received lots of amateur stations with the receiver. Then the receiver was remade by me in general coverage HF- Receiver. I could receive with great quality forbidden BBC, Voice of America lots broadcasting stations and easy found at those times "Numbers Stations." Design of my receiver was not strictly followed by the article. Later I made several such receivers for 19, 16 and 13- meter broadcasting bands. Some people asked me did it for receiving BBC and Voice of America that were strongly jammed at others broadcasting bands.

However, I have seen (and heard its work) this receiver made strictly according to the article. It was in Kharkov, when I studied at the Radio and Electronics University. The receiver was made at the far 50s and still used at one Ukraine Amateur Station as a backup receiver.

The receiver works well with any types of the tubes. It works good at low voltage, as I remember, 25 Volts was enough for the receiver. Nominals of the resistors and capacitors are not critical. Tolerance in 30- 50% is okey.

Figure 1 shows schematic of the receiver. It is classical regenerative receiver with RF Stage and Separate Audio Amplifier.

Radio # 4, 1953
Front Cover

RF Stage (Tube 1) allows use a short wire as antenna. If you would use good long wire you need decrease capacity of the capacitor C5 to 1...- 8- pF. To make general coverage HF Receiver you need use dual variable capacitor to tune input and regenerator inductors. Take attention to the feeding of the heating of the tubes. At the receiver it is used artificial ground to eliminate hum in the phones. Battery operated (or DC – powered) heating would be a good decision.

Title of the Article

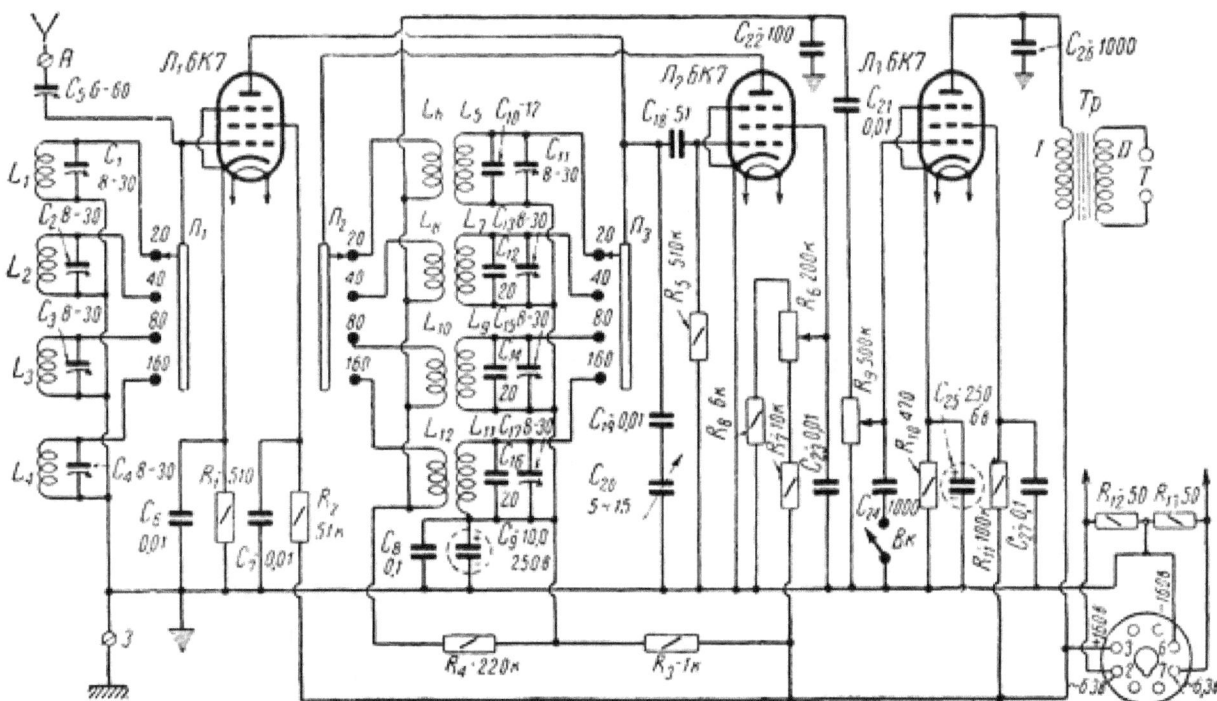

Figure 1 Schematic Diagram of the HF Receiver for Beginner

Capacitor C20 should be with air dielectric. Vernier of any kind would be useful with the capacitor.

Receiver has potentiometer R9 for audio signal. It is useful part of any regenerative receiver where some signals are at the hearing border but another very loud. Capacitor C24 (with help of a switch) grounded first grid of the audio tube. It is decreased noise in the headphones at CW- reception. Any Audio Transformer from an old tube receiver may be used for the receiver.

Figure 2 shows design of the inductors L1… L12. All inductors were wound on empty gun cartridges in diameter 20- mm. Table 1 shows data for the inductors.

Figure 3 shows design of the receiver. At the top of the inductors trimmer capacitors are installed. It is very conveniently for tuning of the receiver at amateurs bands. Figure 4 shows the trimmer capacitors at top of the inductors.

The receiver is easy to tuning and fun for operation. At good parts and inductors made according to the Table 1 the receiver works straight away and at the frequencies close to amateur bands.

Viktor Lomanovich, UA3DH

Picture from 1970s

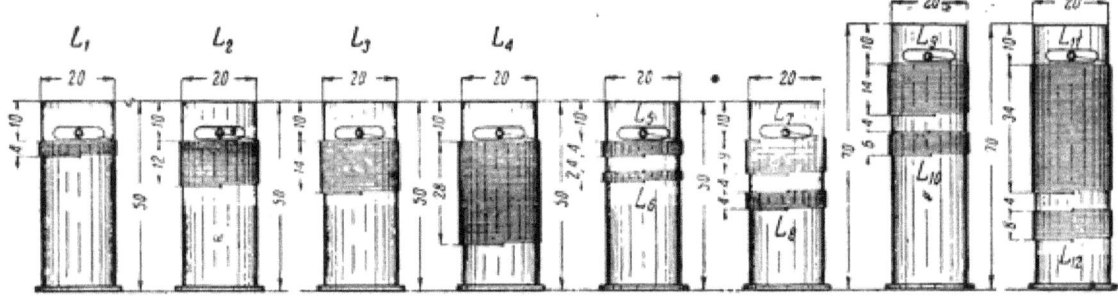

Рис. 2. Эскизы контурных катушек. L₁ содержит 9 витков, а L₂ — 20 витков ПЭЛ 0,5; L₃ — 40 витков, а L₄ — 80 витков ПЭЛ 0,3; L₅ — 9 витков ПЭЛ 0,5; L₆ — 5 витков ПЭЛ 0,3; L₇ — 18 витков ПЭЛ 0,5; L₈ — 14 витков, L₉ — 38 витков, L₁₀ — 15 витков, L₁₁ — 100 витков и L₁₂ — 20 витков ПЭЛ 0,3

Figure 2
Design of the Inductors L1... L12

Table 1 Data for the Inductors.

L1	L2	L3	L4	L5	L6	L7	L8	L9	L10	L11	L12
9 turns Wire 0.5-mm	20 turns Wire 0.5-mm	40 turns Wire 0.3-mm	80 turns Wire 0.3-mm	9 turns Wire 0.5-mm	5 turns Wire 0.3-mm	18 turns Wire 0.5-mm	14 turns Wire 0.3-mm	38 turns Wire 0.3-mm	15 turns Wire 0.3-mm	100 turns Wire 0.3-mm	20 turns Wire 0.3-mm

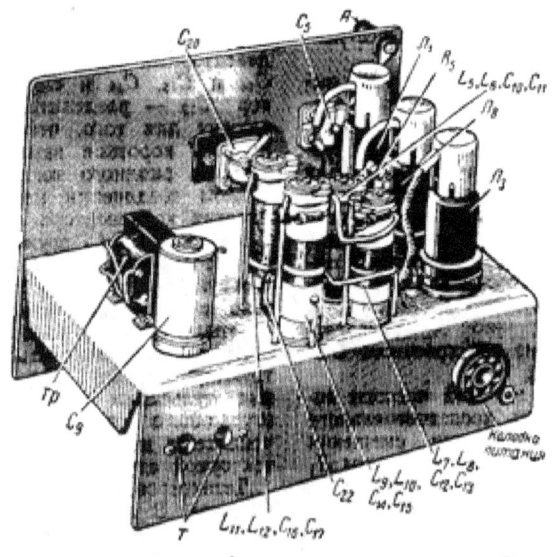

Рис. 4. Вид на шасси приемника сзади

Figure 3 Design of the HF Receiver

Рис. 3. Общий вид контурных катушек с подстроечными конденсаторами: а — катушек L₁, L₂, L₃, L₄; б — остальных катушек

Figure 4 Trimmer Capacitors at Top of the Inductors

FREE e-magazine edited by hams for hams
Devoted to Antennas and Amateur Radio
www.antentop.org

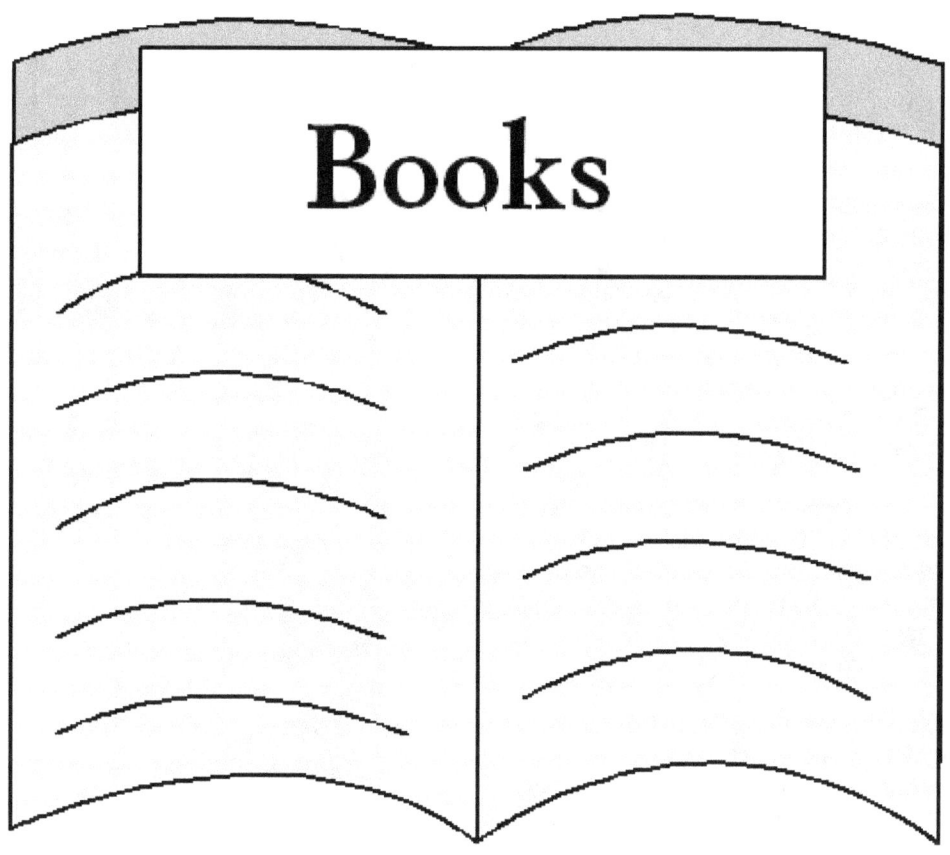

Direct Conversation Technique for Radio Amateurs

Direct Conversation Technique for Radio Amateurs.
Vladimir Polyakov, RA3AAE,
Ph. D in Technical Science
Publishing House: Patriot
Moscow, 1990

Book "Direct Conversation Technique for Radio Amateurs" was published in Moscow in 1990 year. The book was compliance of two early published two books- "Direct Conversation Transceivers" (the book was published in 1984) and "Direct Conversation Receivers" (the book was published in 1981). Lots new stuff there was added to the "Direct Conversation Technique for Radio Amateurs." The book was very popular among radio amateurs of the USSR and then ex- USSR countries. I can say that the book was a Bible Book for Direct Conversation Technique. The book was published in 200, 000 copies so lots amateurs could have access to the book. The book gave way to amateur radio for many beginner hams. As well the book inspired experiments with Direct Conversation Technique.

I have all three books written by RA3AAE. Lots hours were spending with soldering iron and one of the books. The book was found in the internet at public domain (I really do not remember where I found it) and now the book available at ANTENTOP site. Such books cannot be disappeared. It should be serve for Radio Amateurs for long years.

Link to load the book:

http://www.antentop.org/library/shelf_dc_technique.htm

The book was published in Russian. I cannot find the English translation of the book.

73! de va3znw

Direct Conversation Technique for Radio Amateurs

Underground and Ground Antennas

Underground and Ground Antennas.
Georgiy A. Lavrov
Aleksey S. Knyazev
Publishing House: Sovetskoe Radio
Moscow, 1965

Book "Underground and Ground Antennas" was published in Moscow in 1965 year. The book was published in small numbers of printed copies- 6500 e.a. It is a very small for USSR where the textbook was published in 100, 000 and more copies. However it was not usual Antenna textbook. It was Antenna textbook on to classified subject- Underground and Ground Antennas. The antennas were used in military application not for civil using.

Underground and Ground Antennas were used at underground Military Communication Centres that should be survived after nuclear strike. As well later Underground and Ground Antennas were used in secret Soviet System "Perimeter." The system allowed make last nuclear strike ever personnel responsible for the missile were dead.

The system (Perimeter) was some time not operable in full however then was restored. Some Underground and Ground Antennas were described at Antentop pages. Just go to
http://www.antentop.org/book/c_underground.htm

There are you find some information on underground antennas. Though the book did not open existing underground antennas installed in the USSR the theoretical sides of the underground antennas were described in full. As I know it was in the USSR sole book on Underground and Ground Antennas that was in open access. No one book on the theme was published for open access after the book.

The book not only gives wide theoretical knowledge on the antenna thematic. The book has astonished list of the past Antenna Books and articles that could be accessible from usual library.

The book was published in Russian. I cannot find the English translation of the book.

Underground and Ground Antennas

Link to load the book:

http://www.antentop.org/library/shelf_underground_antennas.htm

The book was found in the Internet at one of the Public Domain. Book in Russian.

73! de va3znw

Jones Antenna Handbook

Jones Antenna Handbook contains only 66 pages but information that is given there is invaluable. The book describes in short and strict range of antennas used by amateurs and professionals on HF- VHF- UHF- Bands. The book gives very good view to the amateur history of the far- 30s.

Link to load the book:
http://www.antentop.org/library/shelf_Jones.htm

Contents

Antenna Theory

Antennas for Transmitting

Directional Antennas

Antennas for Receiving

Antennas for Ultra- High Frequencies

Special Purposes Antennas

160 Meter Antennas

Antenna Coupling Systems

Antenna Charts

Measuring Equipment

Introduction.

The aim of The Jones Antenna Handbook is to provide practical guidance in the selection and construction of that type of equipment which is best suited for a specified purpose and location. Because of the great diversity in types and the conflicting opinions as to their relative merits, the reader may well be puzzled in his choice of what is best for his particular needs.

The antenna is a most important factor in determining the performance of a radio transmitter or receiver. The type should be selected on the basis of known facts and not guesswork…

Field Antenna Handbook

It is US NAVY Field Antenna Handbook. As usual the book is written by super simple language and do not crowded with bunch of unnecessary information and words.

Link to load the book:
http://www.antentop.org/library/shelf_field_ant.htm

Contents

Electromagnetic Radiation

Radio Waves

Radio Communication Circuit

Propagation Fundamentals

Noise

Forming a Radio Wave

Radiation

Polarization

Directionality

Resonance

Reception

Reciprocity

Impedance

Bandwidth

Gain

Take- Off Angle

Grounded Antenna Theory

Types of Grounds

Azimuth

And more….

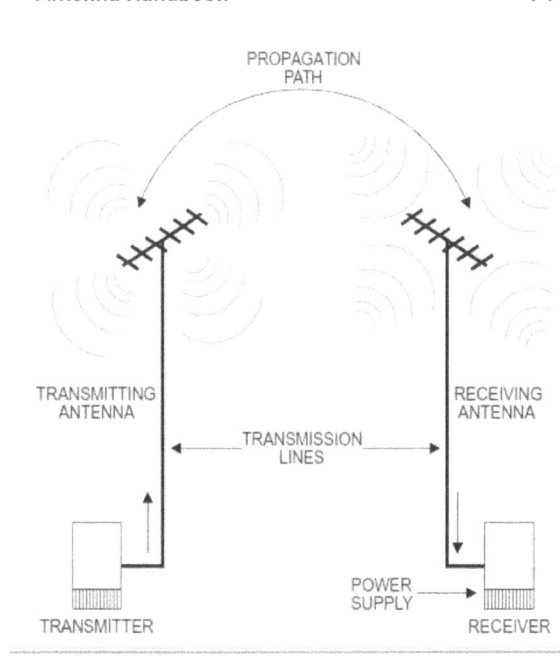

Construction of a Rhombic Receiving Antenna

This is old and reliable book on the subject of the Rhombic Receiving Antennas. The book was published in 1943. No Internet, no empty words were in the times. The book contains just experience and needed information. Personally for me it was very interesting to read the book. Antennas described here works at the 4- 22- MHz. It is possible to approximate the antennas to the other band coverage included of course amateurs HF- Bands.

Link to load the book:
http://www.antentop.org/library/shelf_rhombic_antennas.htm

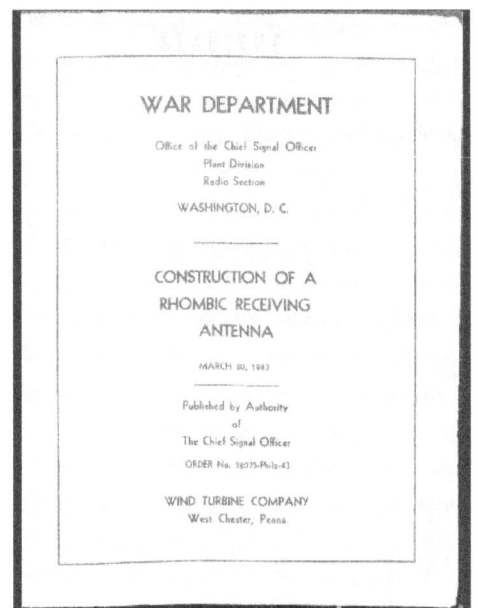

The Book is taken from : http://www.Radionerds.com

http://www.Radionerds.com is a completely free and open resource for radio restoration information. Our goal is information preservation, not control and restrictions. We want you to copy and duplicate these manuals anywhere you wish, the point here is to make these manuals so easy to find on the internet, that no one would bother to buy a CD or try some CIA/MI-6 scheme of authentication to allow you to have something that is PUBLIC DOMAIN. That's right citizens, US Military manuals are all Public Domain!... copy them - all of them... put them on your own site, put them on your friend's site, copy them to a thumb drive and send to your Aunt Minnie, whatever. Selling them or restricting access is *wrong*, that's why we're making it easy for everyone to get them - for free! You can help to keep our site at the top of search engines by linking back to us.

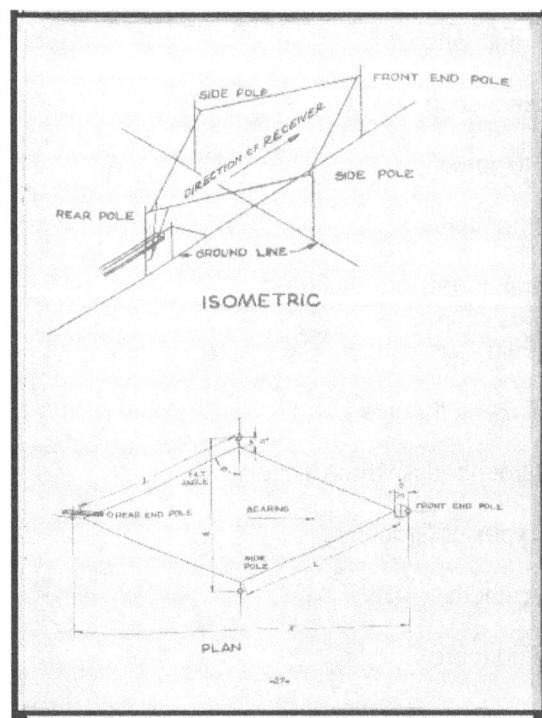

Antennas and Antenna Systems

This is old and reliable book on the subject of the **Antennas and Antenna Systems**. The book was published in 1943. No Internet, no empty words were in the times. The book contains just experience and needed information. Personally for me it was very interesting to read the book. Old good book with interesting antique stuff inside.

Link to load the book:
http://www.antentop.org/library/shelf_antenna_systems.htm

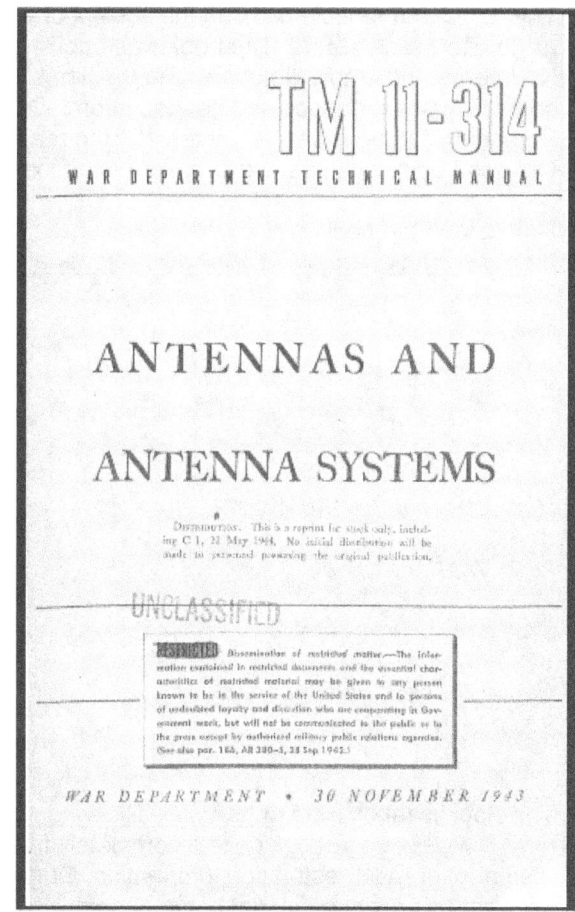

The Book is taken from : http://www.Radionerds.com
http://www.Radionerds.com is a completely free and open resource for radio restoration information. Our goal is information preservation, not control and restrictions. We want you to copy and duplicate these manuals anywhere you wish, the point here is to make these manuals so easy to find on the internet, that no one would bother to buy a CD or try some CIA/MI-6 scheme of authentication to allow you to have something that is PUBLIC DOMAIN.

That's right citizens, US Military manuals are all Public Domain!... copy them - all of them... put them on your own site, put them on your friend's site, copy them to a thumb drive and send to your Aunt Minnie, whatever. Selling them or restricting access is *wrong*, that's why we're making it easy for everyone to get them - for free! You can help to keep our site at the top of search engines by linking back to us.

Antennas and Radio Propagation

This is old and reliable book on the subject of the **Antennas and Antenna Systems**. The book was published in 1953. No Internet, no empty words were in the times. The book contains just experience and needed information. Personally for me it was very interesting to read the book. Old good book with interesting antique stuff inside.

Link to load the book:
http://www.antentop.org/library/radio_propagation.htm

Contents

Introduction

Modes of Propagation

Half Wave and Quarter Wave Antennas

Long Wire Antennas

Driven and Parasitic Arrays

Radio Direction Finding Antennas

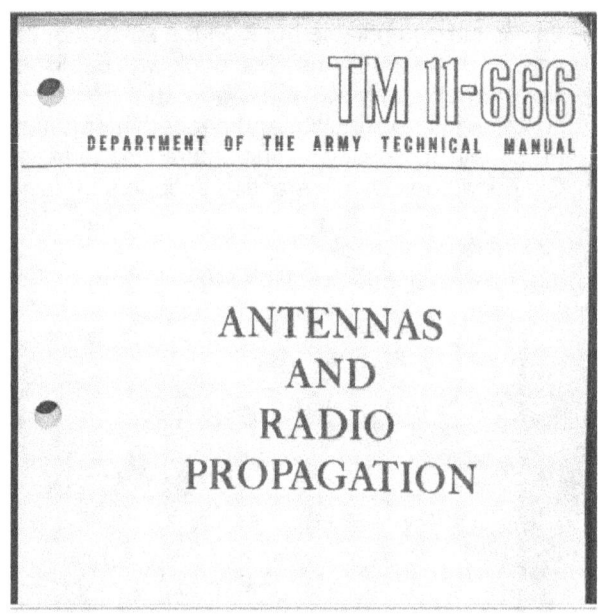

The Book is taken from : http://www.Radionerds.com
http://www.Radionerds.com is a completely free and open resource for radio restoration information. Our goal is information preservation, not control and restrictions. We want you to copy and duplicate these manuals anywhere you wish, the point here is to make these manuals so easy to find on the internet, that no one would bother to buy a CD or try some CIA/MI-6 scheme of authentication to allow you to have something that is PUBLIC DOMAIN.

That's right citizens, US Military manuals are all Public Domain!... copy them - all of them... put them on your own site, put them on your friend's site, copy them to a thumb drive and send to your Aunt Minnie, whatever. Selling them or restricting access is *wrong*, that's why we're making it easy for everyone to get them - for free! You can help to keep our site at the top of search engines by linking back to us.

Design Handbook for High Frequency Radio Communications Systems

This is old and reliable book on the subject of the **Military Handbook: Design Handbook for High Frequency Radio Communications Systems.**

The book was published in 1986. Lots interesting information on Antennas, Propagation, Modulation and so on. Strongly recommended....

Link to load the book:
http://www.antentop.org/library/shelf_design_handbook.htm

The Book is taken from : http://www.Radionerds.com
http://www.Radionerds.com is a completely free and open resource for radio restoration information. Our goal is information preservation, not control and restrictions. We want you to copy and duplicate these manuals anywhere you wish, the point here is to make these manuals so easy to find on the internet, that no one would bother to buy a CD or try some CIA/MI-6 scheme of authentication to allow you to have something that is PUBLIC DOMAIN.

That's right citizens, US Military manuals are all Public Domain!... copy them - all of them... put them on your own site, put them on your friend's site, copy them to a thumb drive and send to your Aunt Minnie, whatever. Selling them or restricting access is *wrong*, that's why we're making it easy for everyone to get them - for free! You can help to keep our site at the top of search engines by linking back to us.

MILITARY HANDBOOK

DESIGN HANDBOOK

FOR

HIGH FREQUENCY RADIO

COMMUNICATIONS SYSTEMS

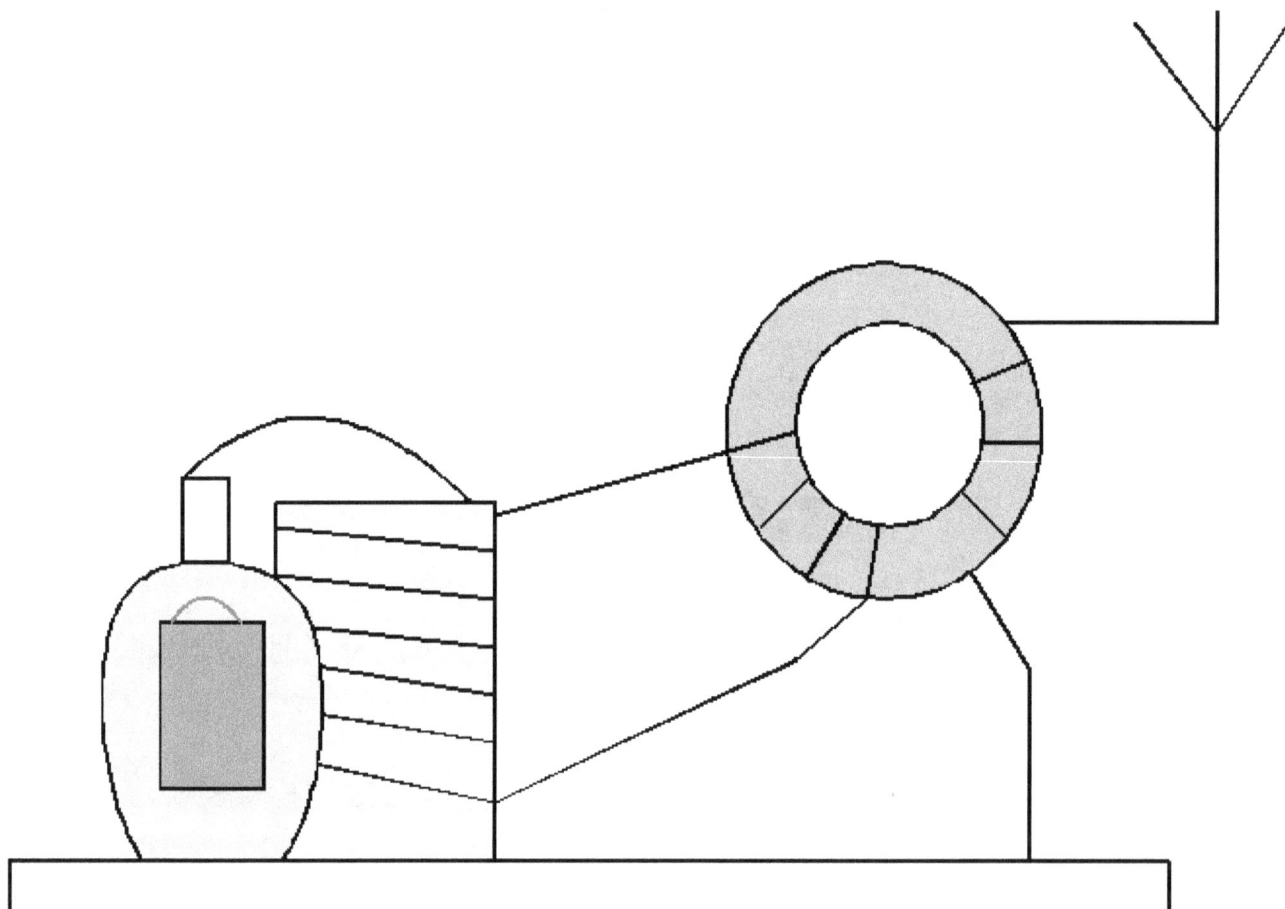

www.antentop.org

Broadband Transformer 50/200- Ohm

By: Sergey Popov, RZ9CJ, Ekaterinburg, Russia

Credit Line: http://qrz-e.ru/forum/29-786-2

Below I describe a simple way to make broadband transformer 50/200 Ohm with isolated windings. (Theoretically the transformer is for 50/140- Ohm. However it works fine for most common using 50/200 Ohm.)

Figure 1 shows schematic of the transformer. At first you need take a ferrite ring (or several identical ferrite rings) with permeability 600… 2000. OD should be 40- 50- mm. Power going through the transformer depends on the sizes. I took two ferrite rings with permeability 2000 for the transformer. Figure 2 shows the rings.

The rings stuck together. Rings should be protected from atmospheric influences. It may be done with protection lacquer or just with electrical insulation tape. Figure 3 shows two rings wrapped with electrical insulation tape. Of course, at the antenna the transformer should be protected from straight atmospheric influences with help of a simple cover.

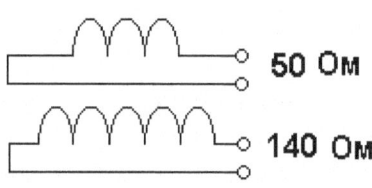

Figure 1 Schematic of the Broadband Transformer 50/200 Ohm

Second winding has two turns around each turn of the first winding. Wires should be close and parallel each to other.

Figure 6 shows beginning of the second winding at one side of the ring. In one turn (first winding) is in brown, two turns (second winding) are in yellow. Figure 7 shows ready second winding.

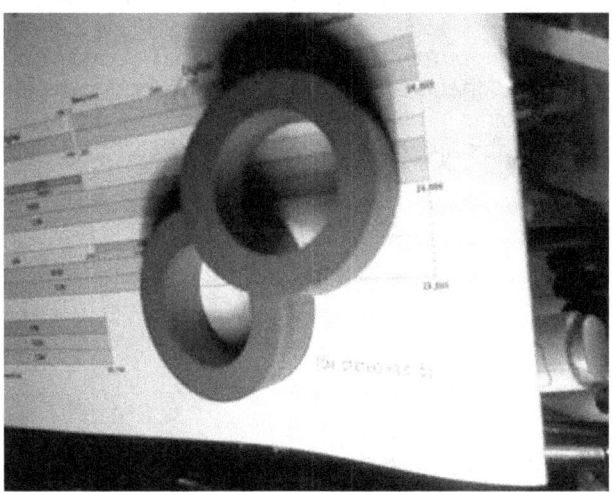

Figure 2 Rings for the Broadband Transformer

Figure 3 Two Rings Wrapped with Electrical Insulation Tape

I used thin wire in diameter near 1.5- mm (14- AWG) for the windings. First winding contains 2 turns. Figure 4 shows transformer with the winding. Then one turns is moved to the opposite side of the ring. It is the first winding. Figure 5 shows the first winding.

First winding is for coaxial cable 50 Ohm the second one for antenna 200 Ohm. After that it is pressed the turns together and fixed it with electrical insulation tape. Transformer 50/200 Ohm is ready. Figure 8 shows the transformer.

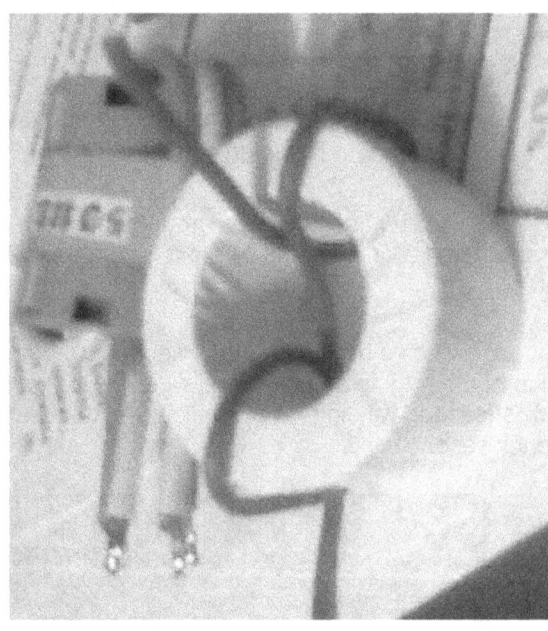

Figure 4 Transformer with 2 turns winding

Figure 7 Ready Second Winding

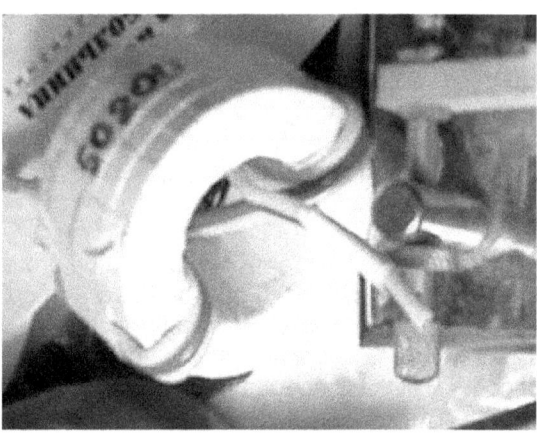

Figure 8 50/200 Ohm Transformer

The transformer may use with an OCF (Windom) Antenna that is shown below. For DX communication feed points should be at up position compare to the long wire of the antenna.

Figure 5 First Winding

Figure 6 Beginning of the Second Winding at One Side of the Ring

Антенна OCF
7 , 14 , 18 , 21 , 24 , 28 мГц

4 м

38,2 м

трансформатор
200/50 Ом

Для DX связей точка запитки должна быть вверху а полотно антенны наклонено вниз

RZ9CJ

Two Broadband Symmetrical Transformers for HF and VHF Bands

By: Alex Karakaptan, UY5ON, ES4UY, Kharkov, Ukraine

Below there are described two broadband transformers with transformation ratio 1:5 and 1:10. First transformer could match 50 Ohm to 250 Ohm the second one could match 50 Ohm to 500 Ohm. The transformers may be used to feed different types of hi- ohmic antennas, for example, G5RV. The transformers provide symmetrical output. The transformers provide SWR 1.15 AT 144 MHz, 1.1 at 70- MHz, 1.0 at 50- MHz, 1.1 at 3.5- MHz and 1.15 at 1.9- MHz. The transformers do not overheated at 500- Watts power going through.

Figure 1 shows design of the transformer with ratio 1: 5 (50/250 Ohm). Main core of the transformer is a ferrite tube taken from an RFI filter that was placed on control wires of old CRT monitor. Such tube may have OD 18… 20- mm, ID 8… 9- mm, and length 25- 28- mm. Permeability of the core is near 800- 1000.

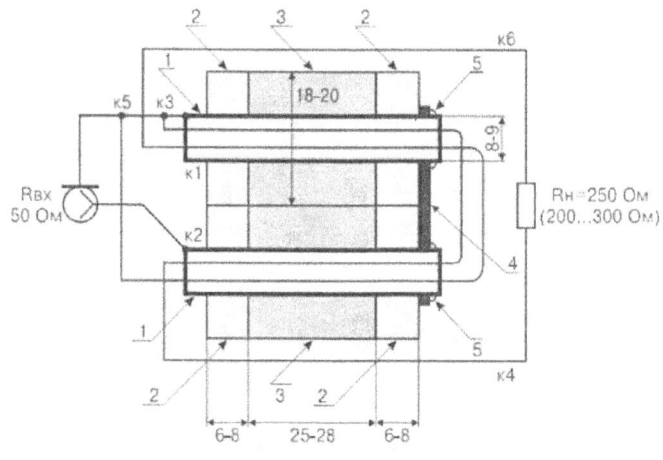

Figure 1 Design of the Transformer with Ratio 1: 5 (50/250 Ohm)

Length ferrite core for the transformer should be 37... 44- mm. So, two ferrite rings with equal to the tube OD and ID placed from the both sides of the core. First winding of the transformer (k1k2 at Figure 1) consist of two copper tubes (pos.1) that are inserted into the ferrite cores. The tubes should have a minimal gap between the cores. The tubes are shorted by a jumper (pos.4). The jumper made of a copper strip. The jumper is soldered (pos.5) to the copper tubes. Second winding (k3k4 + k5k6) made by teflon wire in diameter 1.5- 2.0 mm (15… 12 AWG).

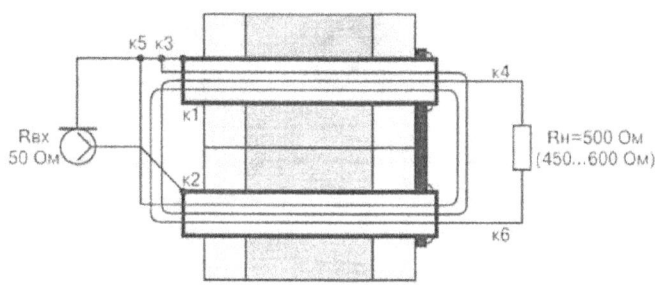

Figure 2 Design of the Transformer with Ratio 1: 10 (50/500 Ohm)

Figure 2 shows design of the transformer with ratio 1: 10 (50/500 Ohm). The transformer has design similar to the transformer with ratio 1: 5 (50/250 Ohm). Difference is only in the second winding that design is cleared from Figure 2.

Both transformers were tested at real antennas. They worked perfect at least much better the usual transformer made on a ferrite ring. The transformers have equal frequency parameters from 1.9 to 144 MHz that could not provide the usual transformers on ferrite ring.

Credit Line: Radio Hobby # 6, 2014

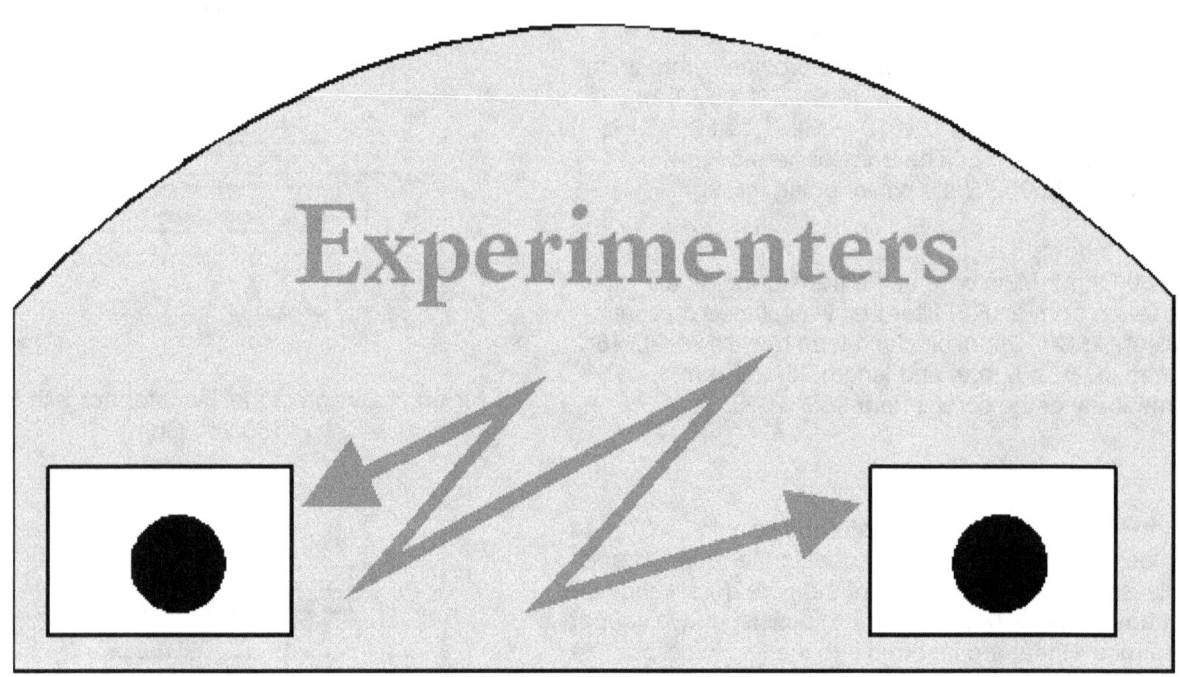

Experimenters with Microwave Oven

Almost everyone has at home a Microwave Oven. It is possible make some experimenters with it. Most interesting and visual experiment is Experiment with Bulbs. We may find how microwaves effect to incandescent (filament) and CFL bulbs.

I bought at nearest Walmart a kit of bulbs for repair of my Christmas-tree light decorates. **Figure 1** shows the kit. The kit contains 5 bulbs for 2.5- 3.0- V. **Figure 2** shows the bulb from the kit. For me it was interesting the behavior of the bulb in my Microwave Oven. I put the lamp in the oven with some meals and turn on the oven. The bulb is flashing very brightly according to oven's magnetron pulse radiation. The lamp is flashing anywhere in the Oven- at the turning plate or being sealed at the Oven's door. Adding several centimeters of wire to the bulb leads may kill the bulb in the oven. Leads of the bulbs may be soldered together at the ends. It is not prevent the Lamp from flashing in the working microwave.

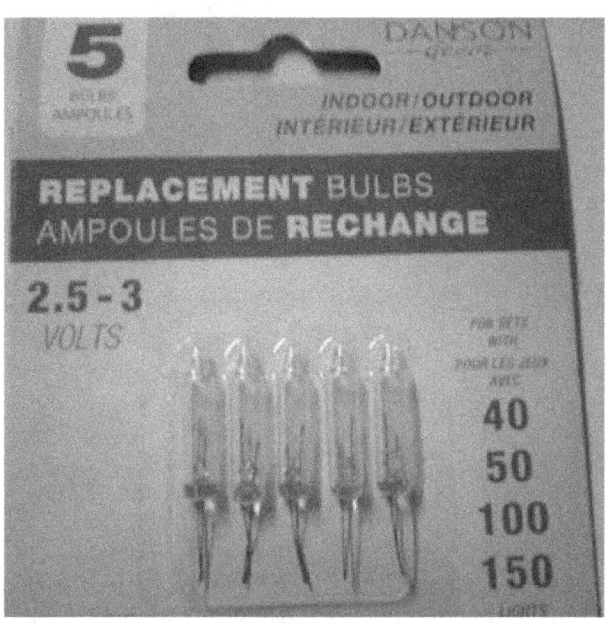

Figure 1 Kit for repair Christmas-tree light decorates

Figure 2 Bulb from the kit for repair Christmas-tree light decorates

Next subject for experiment was a 120V/20W filament bulb. It was lamp from an old microwave. Bare lamp did not flashing in the microwave Oven. Then I added to the lamp leads a loop in perimeter 12 centimeters. The loop was a loop antenna that tuned to 2.45- GHz. The lamp was a load for the antenna. Just remind, that oven's magnetron works at the 2.45- GHz. **Figure 3** shows design of the probe. I put the probe inside of the Microwave Oven with some meals and turn on the oven.

Figure 3 20W Probe (A)

The probe bulb is flashing very brightly according to Oven's magnetron pulse radiation. I experimented with filament bulb for 120V/40W, 120V/60W and 120V/100W. All bulbs flashing enough brightly in depend of a load (water, meals, etc) that was being inside of the Oven.

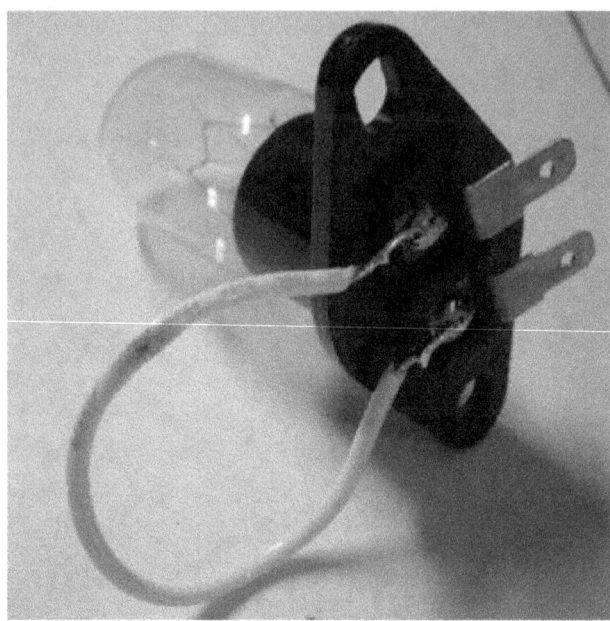

Figure 3 20W Probe (B)

Next step was to test a CFL lamp. I have lots CFL lamp with removed base (I used the base to find electronics parts for my experiments). Figure 4 shows the CFL lamp with removed base. I put the CFL lamp inside of the Microwave Oven with some meals and turn on the Oven. The CFL lamp is flashing steady very brightly do not depend on to oven's magnetron pulse.

It was understandable for me. CFL lamp makes light because of luminophor on the inner side of tube. The luminophor has low reaction - turn off any CFL lamp and see that it was not be dark immediately. At usual conditions the CFL lamp is lighting because of low temperature plasm inside of the lamp. The plasm generate UV that luminophor transform into visible light. In normal working conditions the plasm occurs due to high voltage across the CFL bulb leads. In the Microwave Oven the plasm occurs due microwave radiation. I tested in the Microwave Oven working CFL with working base. It is not damaged the lamp.

Figure 5 shows blowing a CFL Lamp inside a Microwave Oven. Figure 6 shows blowing a small lamp inside a Microwave Oven.

73! Igor Grigorov

Figure 4 CFL Lamps

Figure5 Blowing a CFL Lamp inside a Microwave Oven

Figure 6 Blowing a small lamp inside a Microwave Oven

ANTENTOP

ANTENTOP is *FREE e- magazine*, made in **PDF**, devoted to antennas and amateur radio. Everyone may share his experience with others hams on the pages. Your opinions and articles are published without any changes, as I know, every your word has the mean.

A little note, I am not native English, so, of course, there are some sentence and grammatical mistakes there... Please, be indulgent!

Publishing: If you have something for share with your friends, and if you want to do it *FREE*, just send me an email. Also, if you want to offer for publishing any stuff from your website, you are welcome!

Copyright: Here, at ANTENTOP, we just follow traditions of *FREE* flow of information in our great radio hobby around the world. A whole issue of ANTENTOP may be photocopied, printed, pasted onto websites. We don't want to control this process. It comes from all of us, and thus it belongs to all of us. This doesn't mean that there are no copyrights. There is! Any work is copyrighted by the author. All rights to a particular work are reserved by the author.

Copyright Note: Dear friends, please, note, I respect Copyright. Always, when I want to use some stuff for ANTENTOP, I ask owners about it. But... sometimes my efforts have no success. I have some very interesting stuff from closed websites however their owners keep silence... as well as I have no response on some my emails from some owners

I have a big collection of pictures. I have got the pictures and stuff in different ways, from *FREE websites*, from commercial CDs, intended for *FREE using*, and so on... I use to the pictures (and seldom, some stuff from free and closed websites) in ANTENTOP. *If the owners of the Copyright stuff have concern*, please, contact with me, I immediately remove any Copyright stuff, or, if it is necessary, all needed references will be made there.

Business Advertising: *ANTENTOP is not a commercial magazine.* Authors and I (Igor Grigorov, the editor of the magazine) do not get any profit from any issue. But of course, I do not mention from commercial ads in ANTENTOP. It allows me to do the magazine in most great way, allows me to pay some money for authors to compensate their hard work.

So, if you want paste a commercial advertisement in ANTENTOP, please contact me.

Book Advertising: I believe that *Book Advertising* is a noncommercial advertisement. So, Book Advertising is *FREE* at ANTENTOP. Contact with me for details.

And, of course, tradition approach to ANY stuff of the magazine:

BEWARE:

All the information you find at *AntenTop website* and any hard (printed) copy of the *AnTentop Publications* are only for educational and/or private use! I and/or authors of the *AntenTop e- magazine* are not responsible for everything including disasters/deaths coming from the usage of the data/info given at *AntenTop website/hard (printed) copy of the magazine*.

You use all these information of your own risk.

---

www.ingramcontent.com/pod-product-compliance
Lightning Source LLC
Chambersburg PA
CBHW080930170526
45158CB00008B/2229